Advances in 21st Century Human Settlements

Series Editor

Bharat Dahiya⓲, School of Global Studies, Thammasat University, Bangkok, Thailand

Editorial Board

Andrew Kirby, Arizona State University, Tempe, USA

Erhard Friedberg, Sciences Po-Paris, Paris, France

Rana P. B. Singh, Banaras Hindu University, Varanasi, India

Kongjian Yu, Peking University, Beijing, China

Mohamed El Sioufi, Monash University, Clayton, Kenya

Tim Campbell, Woodrow Wilson Center, Washington, USA

Yoshitsugu Hayashi, Chubu University, Kasugai, Japan

Xuemei Bai, Australian National University, Canberra, Australia

Dagmar Haase, Humboldt University, Berlin, Germany

Ben C. Arimah, United Nations Human Settlements Programme, Nairobi, Kenya

Indexed by SCOPUS

This Series focuses on the entire spectrum of human settlements – from rural to urban, in different regions of the world, with questions such as: What factors cause and guide the process of change in human settlements from rural to urban in character, from hamlets and villages to towns, cities and megacities? Is this process different across time and space, how and why? Is there a future for rural life? Is it possible or not to have industrial development in rural settlements, and how? Why does 'urban shrinkage' occur? Are the rural areas urbanizing or is that urban areas are undergoing 'ruralisation' (in form of underserviced slums)? What are the challenges faced by 'mega urban regions', and how they can be/are being addressed? What drives economic dynamism in human settlements? Is the urban-based economic growth paradigm the only answer to the quest for sustainable development, or is there an urgent need to balance between economic growth on one hand and ecosystem restoration and conservation on the other – for the future sustainability of human habitats? How and what new technology is helping to achieve sustainable development in human settlements? What sort of changes in the current planning, management and governance of human settlements are needed to face the changing environment including the climate and increasing disaster risks? What is the uniqueness of the new 'socio-cultural spaces' that emerge in human settlements, and how they change over time? As rural settlements become urban, are the new 'urban spaces' resulting in the loss of rural life and 'socio-cultural spaces'? What is leading the preservation of rural 'socio-cultural spaces' within the urbanizing world, and how? What is the emerging nature of the rural-urban interface, and what factors influence it? What are the emerging perspectives that help understand the human-environment-culture complex through the study of human settlements and the related ecosystems, and how do they transform our understanding of cultural landscapes and 'waterscapes' in the 21st Century? What else is and/or likely to be new vis-à-vis human settlements – now and in the future? The Series, therefore, welcomes contributions with fresh cognitive perspectives to understand the new and emerging realities of the 21st Century human settlements. Such perspectives will include a multidisciplinary analysis, constituting of the demographic, spatio-economic, environmental, technological, and planning, management and governance lenses.

If you are interested in submitting a proposal for this series, please contact the Series Editor, or the Publishing Editor:
Bharat Dahiya (bharatdahiya@gmail.com) or
Loyola D'Silva (loyola.dsilva@springer.com)

Daniel Hoornweg

Canada's Cities in a Changing World 1920–2120

The Halftime Report

Daniel Hoornweg
Ontario Tech University
Oshawa, ON, Canada

ISSN 2198-2546 ISSN 2198-2554 (electronic)
Advances in 21st Century Human Settlements
ISBN 978-981-96-7932-4 ISBN 978-981-96-7933-1 (eBook)
https://doi.org/10.1007/978-981-96-7933-1

© The Editor(s) (if applicable) and The Author(s) 2025. This book is an open access publication.

Open Access This book is licensed under the terms of the Creative Commons Attribution-NonCommercial-NoDerivatives 4.0 International License (http://creativecommons.org/licenses/by-nc-nd/4.0/), which permits any noncommercial use, sharing, distribution and reproduction in any medium or format, as long as you give appropriate credit to the original author(s) and the source, provide a link to the Creative Commons license and indicate if you modified the licensed material. You do not have permission under this license to share adapted material derived from this book or parts of it.
The images or other third party material in this book are included in the book's Creative Commons license, unless indicated otherwise in a credit line to the material. If material is not included in the book's Creative Commons license and your intended use is not permitted by statutory regulation or exceeds the permitted use, you will need to obtain permission directly from the copyright holder.
This work is subject to copyright. All commercial rights are reserved by the author(s), whether the whole or part of the material is concerned, specifically the rights of translation, reprinting, reuse of illustrations, recitation, broadcasting, reproduction on microfilms or in any other physical way, and transmission or information storage and retrieval, electronic adaptation, computer software, or by similar or dissimilar methodology now known or hereafter developed. Regarding these commercial rights a non-exclusive license has been granted to the publisher.
The use of general descriptive names, registered names, trademarks, service marks, etc. in this publication does not imply, even in the absence of a specific statement, that such names are exempt from the relevant protective laws and regulations and therefore free for general use.
The publisher, the authors and the editors are safe to assume that the advice and information in this book are believed to be true and accurate at the date of publication. Neither the publisher nor the authors or the editors give a warranty, expressed or implied, with respect to the material contained herein or for any errors or omissions that may have been made. The publisher remains neutral with regard to jurisdictional claims in published maps and institutional affiliations.

This Springer imprint is published by the registered company Springer Nature Singapore Pte Ltd.
The registered company address is: 152 Beach Road, #21-01/04 Gateway East, Singapore 189721, Singapore

If disposing of this product, please recycle the paper.

Canada's Largest Cities (by population, Census Metropolitan Areas, CMAs)
Census metropolitan areas (CMA, 1 July 2024 population): **St. John's** 239,316, **Halifax** 530,167, Moncton 188,036, Saint John 142,433, Fredericton 122,500, Saguenay 170,380, **Québec** 900,343, **Sherbrooke** 243,517, Trois-Rivières 173,288, Drummondville 108,291, **Montréal** 4,615,154, **Ottawa—Gatineau** 1,660,269, Kingston 192,389, Belleville—Quinte West 123,572, Peterborough 148,027, **Oshawa** 482,359, **Toronto** 7,106,379, **Hamilton** 860,266, **St. Catharines—Niagara** 492,480, **Kitchener—Cambridge—Waterloo** 696,417, Brantford 173,530, Guelph 183,224, **London** 626,260, **Windsor** 483,556, **Barrie** 245,586, Greater Sudbury 191,902, Thunder Bay 133,063, **Winnipeg** 941,641, **Regina** 282,032, **Saskatoon** 367,336, Lethbridge 139,844, **Calgary** 1,778,881, Red Deer 112,759, **Edmonton** 1,631,614, **Kelowna** 251,756, Kamloops 128,233, Chilliwack 126,850, **Abbotsford—Mission** 220,786, **Vancouver** 3,108,941, **Victoria** 441,491, Nanaimo 128,371
Canada total population 41,288,599; all CMAs and census agglomerations 34,958,433; all CMAs 30,893,239 (42 CMAs, **23** > 200 K). From https://www.statcan.gc.ca/en/start (accessed 23/3/2025)

Calgary

Calgary

Edmonton

Halifax

Montreal

Ottawa

Quebec City

Quebec City

Toronto

Vancouver

Winnipeg

This book is dedicated to Abha Joshi-Ghani (July 22, 1959–September 23, 2021). At the World Bank, she received the Excellence in Leadership Award in 2016. I was lucky to have her as my boss and to call her my friend.

She was especially hopeful for the education of girls. Her charity of choice was EDUGirls and the Vimukti Girls School in Rajasthan.

Working with Abha, I was struck by how much better things could be if women were treated fairly. Even in a country like Canada, a man murdered 14 women in Montreal and injured another 14, simply because they wanted to be engineers. Across the country, as many as 4000 Indigenous women are missing or murdered. The world can be hard, especially for women. Abha's indomitable optimism and pragmatism, and her example, remains sorely needed.

Series Editor's Foreword

Canada's Cities in a Changing World 1920–2120: The Halftime Report

In the hustle and bustle of everyday life in cities, reflecting on where things originated and where they might be going is useful. This timely and significant monograph by Dr. Daniel Hoornweg, Associate Professor in the Faculty of Engineering and Applied Science at Ontario Tech University, takes a 200-year timeframe, from 1920 to 2120 AD (about seven generations), to delve into the past, present and future of Canada's cities. Looking back since 1920 and looking forward until 2120, the book provides a 'Half-time Report' on Canada's cities in the first quarter of the 21st century. Any analysis of Canadian cities quickly illustrates how the world's communities are interconnected. This is especially true for Canadian cities as they are home to one of the world's largest shares of immigrants, a trend that is set to increase in the next century (2020 to 2120 AD) as global (urban) demographics shift [1, 2], and the number of climate refugees grows [3–5].

This 'half-time report' discusses the unique strengths and dire challenges of urban Canada and comes at pivotal moment in Canada's and the world's history. Canada's cities are among the world's most energy- and material-intensive; i.e. their energy and material consumption is one of the highest in the world. They will need to lead the shift from the *great acceleration* experienced over the last 100 years, to the *great deceleration* now growing in impetus. Canada's wealth, largely derived from resource development, is shifting (too slowly) to a low carbon, more circular and more equal economy. Cities in Canada, and the rest of the world, are leading this historic shift.

Canada's cities provided many firsts. For example, Ottawa closed streets to cars in 1970, encouraging bicycles on Sunday. A lesser-known fact is that this was replicated and expanded by cities like Bogotá that launched *Ciclovía* in 1974 [6]. The world's first business improvement area (BIA) started in Toronto's Bloor Street West in 1970 [7]. There are now more than 6,000 BIAs around the world [8].

Canada's cities are highly fractured, with very few powers as they operate within a constitutional framework that gives enormous authority to the provinces. Canada's largest city, Toronto Region, the natural urban system in need of service planning and optimization, is made up of 106 local (municipal) governments with 1,200 municipal politicians, 34 transit agencies, 17 electricity distributors, 25 school boards, 8 health networks, 70 chambers of commerce and 25 publicly funded post-secondary institutions with more than 40 campuses. Toronto Region makes up 70 per cent Ontario's economy and almost a third of Canada's, yet there is not a professional employee or politician, at any level, who speaks for the whole community. Montreal is similarly splintered, made up of 82 municipalities.

The Toronto city region as a whole is largely at the mercy of its constituent parts. Traffic congestion in Toronto, the worst in North America, and third worst in the world [9], is a powerful example of too many cooks in the kitchen. The resulting obstruction costs more than CAD 11 billion annually in productivity losses alone. Despite considerable analysis, improvements are minimal as the fractured urban service fails to have a systems approach applied.

The recent 'Taylor Swift–The Eras Tour' [10] also provides an example of Toronto's challenges. In an upcoming paper, Hoornweg et al. (2025, in preparation) [11] the 20-country, 52-city, 152-show tour generated about US $3 billion in economic activity. Of the 52 cities, Toronto fared the poorest in terms of economic contribution from the tour. The six shows in Toronto (second only to London's 8 concerts) included 290,000 guests with total revenues about $230 million. The total tax revenues from these expenditures were about $40 million, of which the City of Toronto received about US $3.1 million (8.6%, the rest paid to province and national government; the city also incurred a roughly US $2 million policing and emergency management staffing overtime cost). Compare this to Chicago, that hosted 3 concerts, where the city received US $21 million in tax revenue (about 36%).

Since 1920, 122 new countries emerged or re-emerged from colonial rule. Only 28 countries participated in the 1912 Olympic Games [12], while 204 countries took part in the last Olympics held in Paris in 2024. Many countries fractured over the last 100 years, while Canada and the USA are unique in that despite regional tensions, they consolidated territories.

Populations around the world changed dramatically; however, the number of cities stayed roughly the same, many being the traditional communities of Indigenous peoples who were there long before countries emerged.

Cities scale. This is their superpower. Double the size of a city and its economy more than doubles (~1.15 times), and this can be done for less than twice the infrastructure cost (~0.85 times) [13, 14]. Countries and businesses do not scale. This is a reason why the average lifespan of a business is less than 40 years, and in every country, the main cities are older than the country.

Cities cannot move out of harm's way. This can be a weakness but is also a source of strength. Cities know that they need to work together to defend themselves. Most countries emerge when a group of cities comes together. *Cogitando et Agendo Ducemus*, by thinking *and doing*, we shall lead. Cities bring leadership through pragmatism.

Cities are the closest human construct to the natural world, behaving as complex emerging systems akin to a forest ecosystem or a beehive. The day-to-day governance of a city is usually less partisan than that of a country. Cities emerge naturally when enough people live in the same location. Countries, on the other hand, need narratives, shared beliefs, and common values. Countries issue passports and currency, and leaders spend much of their time trying to win hearts and minds or suppressing dissent. You need passion and a storyline to be able to convince the young to join the military or the rich to pay more taxes.

Countries are only able to project power when their economy is anchored to flourishing cities. A new 'League of Communities' needs to emerge, perhaps modelled after the old Hanseatic League of Cities [15, 16] that brought together cities in a protective network that facilitated trading and support between communities. The League of Communities might start with Canadian cities but needs to quickly include communities across the world as the 21st century's big challenges are truly global.

Future generations need representation. In the next 50 years, millions of people are expected to move to cities in the Global North. Migration is fast becoming a key component of sustainability [17]. As cities in the Global North grow, they need to do so in a way that causes minimal impact on local and global earth systems while continuing to accommodate future generations. As demographics shift, successful cities and their countries need to be able to attract and accommodate migrants.

Evoking Freud's [18] *Civilization and Its Discontents*, the author argues that the fundamental paradox of civilization (i.e. cities and communities) is that civilization is a means to protect us from unhappiness, and yet it is our largest source of unhappiness. The common good is predicated on constructive relationships with one another and between cities. Freud acknowledged the ill will within many hearts and said that civilization exists to restrain these impulses. The new 'League of Communities' will recognize that this same ill will may well exist between communities; however, communities need to exist within a common civilization.

If the ill will of humans and nations is allowed to progress unencumbered, planetary boundaries and systems will be overrun [19, 20], bringing about the collapse of major parts of today's civilization. Bringing together a civic community that complements the existing Westphalia system of nations [21, 22] provides the potential to add a voice of sufficient clarity and credibility to temper the ill wills of people and their nations.

The book concludes with a call to action. Recognizing the rapid decline of Canada's wellbeing—dropping from 5th in the 2015 World Happiness Report [23] to 18th place in the one for 2025 [24], especially for Canada's youth (among the largest disparity between youth and old cohorts).

In addition to urging Canadian cities to take on a catalytic role in a 'League of Communities', the book recommends two urban-based narratives: integrating migration and sustainability, and balancing scarcity and sufficiency. Communities need to nurture a metropolitan mindset, or even better, a sustainability mindset [25].

Canadian cities, through pragmatic initiatives like support to international students, and their home countries, and lowering the cost of sending remittances, can quickly build international support. By providing annual sustainability data for urban areas of at least Canada's larger communities—for instance, census metropolitan areas (CMA), and possibly providing a share of value-added taxes back to the specific community of generation (read: CMA), Canada's federal government could also quickly catalyse an emphasis on shared sustainability.

This book highlights that building a hospitable Canada for children being born today, in a world that will be safe and flourishing in 2120 AD is a daunting task. Toronto was first settled by the Iroquois in the village of Teiaiagon [26] near today's Humber River. The Constitution of the Iroquois Nation, believed to date back to 1142 when a total eclipse occurred in the region, recommends consideration for seven future generations (about 200 years) in decision-making.

This half-time report for seven generations starting in 1920 AD would be mixed. Degradation of planetary systems is severe, and peaceful coexistence within Canada and around the world appears to be declining. Trends suggest growing conflict and crossing of several major tipping points in the next few decades.

On the other hand, there is cause for optimism. Global populations will peak and markedly decline within the next 100 years. The transition will be difficult, but planetary systems can remediate, and there are more than enough resources, energy and goodwill in the world to provide a high quality of life for all. Cities need to lead this transition. Canadian cities are well-placed to be part of this national and international civic-minded effort.

In closing, I would like to congratulate Daniel for accomplishing the formidable task of preparing this 'Half-time Report' on Canada's cities. I first met him in 2002 at the World Bank headquarters, Washington DC, where we both were part of the Bank's 'Urban Environment Thematic Group'. Daniel is a world-class professional and academic and specializes in energy systems, urban systems and sustainability. He has a distinguished background, including nearly 20 years as Lead Advisor on Sustainable Cities and Climate Change at the World Bank, where he worked with over 350 local governments worldwide. Daniel also served as Chief Safety and Risk Officer, with the Technical Standards and Safety Authority, for the Province of Ontario from 2012 to 2020. His research focuses on urban sustainability, energy use, waste management and climate change, and he is widely published in these fields [27–35]. Additionally, he is co-owner of Querencia Partners Canada Ltd., a consultancy focused on sustainable infrastructure.

Each chapter of the 'Half-time Report' on Canada's cities discusses a specific problem and then provides thought-provoking perspectives on how that problem may be addressed, now and in future. The book is loaded with data and cites important and relevant literature to build arguments around the topics that it explores followed by the messages that Daniel wants to convey to the readers within Canada and beyond.

An innovative feature of this monograph is the inclusion of the chapter titled 'Thoughts of Others', the first part of which is based on the Daniel's 'interviews ... with about 30 people active in the urban sector, half in Canada and half international'. In many ways, this brings in the distinct and diverse views and perspectives of these urban experts on the thematic focus of the book and ground-truths the issues discussed by the author.

Series Editor's Foreword　　　　　　　　　　　　　　　　　　　　　　　　　　xix

This historic monograph provides rich and useful insights into the governance and management of Canada's cities, which will be useful for scholars, researchers, policy- and decision-makers as well as practitioners of sustainable and resilient urban development.

I hope that the readers enjoy this wonderful book as much as I did.

Bharat Dahiya, M.A., M.Plan., Ph.D.
Director
Research Center for Sustainable
Development and Innovation
School of Global Studies
Thammasat University
Bangkok, Thailand

Extraordinary Professor
School of Public Leadership
Stellenbosch University
Stellenbosch
Western Cape, South Africa

References

1. Haase D, Güneralp B, Dahiya B, Bai X, Elmqvist T (2018) Global urbanization: perspectives and trends. In: Elmqvist T, Bai X, Frantzeskaki N, Griffith C, Maddox D, McPhearson T, Parnell S, Romero-Lankao P, Simon D, Watkins M (eds) Urban planet: knowledge towards sustainable cities, Cambridge University Press, Cambridge, pp 19–44
2. United Nations (2024) World population prospects 2024: summary of results. Department of Economic and Social Affairs, Population Division, New York
3. Clement V, Rigaud KK, de Sherbinin A, Jones B, Adamo S, Schewe J, Sadiq N, Shabahat E (2021) Groundswell Part 2: Acting on Internal Climate Migration. © World Bank. http://hdl.handle.net/10986/36248 License: CC BY 3.0 IGO
4. Rigaud KK, de Sherbinin A, Jones B, Bergmann J, Clement V, Ober K, Schewe J, Adamo S, McCusker B, Heuser S, Midgley A (2018) Groundswell: Preparing for Internal Climate Migration. © World Bank. http://hdl.handle.net/10986/29461 License: CC BY 3.0 IGO
5. UN-Habitat (2024) World cities report 2024: cities and climate action. UN-Habitat, Nairobi
6. World Economic Forum (2024) Ciclovía at 50: What we can learn from Bogotá's open streets initiative. World Economic Forum. Authored by Marcela Guerrero Casas. Retrieved from https://www.weforum.org/stories/2024/11/50-years-ciclovia-open-streets-cycling-cars/. Accessed 2 June 2025
7. Bloor West Village (2025) The History. Bloor West Village. Retrieved from https://www.bloorwestvillagebia.com/about/the-first-bia-in-canada/. Accessed 6 July 2025
8. Charenko M (2015) A historical assessment of the world's first Business Improvement Area (BIA): The case of toronto's bloor west village. Canadian Urban Res 24(2):1–19. http://www.jstor.org/stable/26195289

9. Global News (2024) Toronto among world's worst cities for congestion, according to a new report. Global News. Authored by Isaac Callan. https://globalnews.ca/news/10220758/toronto-traffic-world-worst-3rd/. Accessed 2 June 2025
10. City of Toronto (2024) Taylor Swift—The Eras Tour. City of Toronto. 31 October. Retrieved from https://www.toronto.ca/news/taylor-swift-the-eras-tour-torontos-version/. Accessed 6 July 2025
11. Hoornweg D et al (2025) A sustainability assessment of the Taylor Swift Eras Tour (in preparation)
12. International Olympic Committee (2025) Stockholm 1912. International Olympic Committee. Retrieved from https://www.olympics.com/en/news/stockholm-1912. Accessed 6 July 2025
13. Bettencourt LM, Lobo J, Helbing D, Kühnert C, West GB (2007) Growth, innovation, scaling, and the pace of life in cities. Proc Natl Acad Sci 104(17):7301–7306
14. West G (2017) Scale: The universal laws of life, growth, and death in organisms, cities, and companies. Penguin
15. Cramer FH (1949) The Hanseatic League. Current History 17(96):84–89. http://www.jstor.org/stable/45309337
16. Hibbert, AB (2025) "Hanseatic League". Encyclopedia Britannica, 9 April. Retrieved from https://www.britannica.com/topic/Hanseatic-League. Accessed 6 July 2025
17. Adger WN, Fransen S, Safra de Campos R, Clark WC (2024) Migration and sustainable development. Proc Natl Acad Sci USA. 121(3):e2206193121. doi:https://doi.org/10.1073/pnas.2206193121
18. Freud S (1930) Civilization and its discontents (Translated by Joan Riviere). Hogarth Press, London. Retrieved from https://archive.org/details/in.ernet.dli.2015.221667. Accessed 28 June 2025
19. Richardson K et al (2023) Earth beyond six of nine planetary boundaries. Sci Adv 9:eadh2458. https://www.science.org/doi/10.1126/sciadv.adh2458
20. Rockström J, Donges JF, Fetzer I et al (2024) Planetary boundaries guide humanity's future on earth. Nat Rev Earth Environ 5:773–788. doi:https://doi.org/10.1038/s43017-024-00597-z
21. Oxford Reference (2025) Westphalian state system (Quick Reference). Retrieved from https://www.oxfordreference.com/display/10.1093/oi/authority.20110803121924198. Accessed 7 July 2025
22. The Editors of Encyclopaedia Britannica (2025) "Peace of Westphalia". Encyclopedia Britannica, 16 May. Retrieved from https://www.britannica.com/event/Peace-of-Westphalia. Accessed 7 July 2025
23. Helliwell JF, Richard L, Jeffrey S (eds) (2015) World happiness report 2015. Sustainable development solutions network, New York
24. Helliwell JF, Layard R, Sachs JD, De Neve J-E, Aknin LB, Wang S (eds) (2025) World happiness report 2025. Wellbeing Research Centre, University of Oxford
25. Rimanoczy I (2021) The sustainability mindset principles a guide to developing a mindset for a better world. Routledge, London

26. Robertson D (2023) 'Teiaiagon: A village on the west branch of the toronto carrying place. In: von Bitter R, Williamson RF (eds) The history and archaeology of the Iroquois du Nord. Mercury Series, Archaeology Paper 182. Canadian Museum of History and University of Ottawa Press, Ottawa, pp 129–140
27. Hoornweg D (2016) Cities and sustainability: a new approach. Routledge Studies in Sustainability series, Routledge, New York
28. Hoornweg D, Pope K (2017) Population predictions for the world's largest cities in the 21st century. Environ Urbanization 29(1):195–216. doi:https://doi.org/10.1177/0956247816663557
29. Hoornweg D, Hosseini M, Kennedy C, Behdadi A (2016) An urban approach to planetary boundaries. Ambio 45(5):567–580. doi:https://doi.org/10.1007/s13280-016-0764-y
30. Hoornweg D, Freire M, Baker-Gallegos J, Saldivar-Sali A (eds) (2013a) Building sustainability in an urbanizing world: a partnership report. Urban Development Series Knowledge Papers; No. 17. World Bank, Washington DC. http://hdl.handle.net/10986/18665
31. Hoornweg D, Bhada-Tata P, Kennedy C (2013b) Environment: Waste production must peak this century. Nature 502(7473):615–617. doi:https://doi.org/10.1038/502615a
32. Hoornweg D, Bhada-Tata P (2012) What a waste: a global review of solid waste management. Urban Development Series; Knowledge Papers No. 15. World Bank, Washington DC. http://hdl.handle.net/10986/17388
33. Hoornweg D, Freire M, Lee MJ, Bhada-Tata P, Yuen B (2011a) Cities and climate change: responding to an urgent agenda. Urban Development Series, World Bank, Washington DC. http://hdl.handle.net/10986/2312
34. Hoornweg D, Sugar L, Trejos Gómez CL (2011b) Cities and greenhouse gas emissions: moving forward. Environ Urbanization 23:207–227. doi:https://doi.org/10.1177/0956247810392270
35. Dickson E, Baker JL, Hoornweg, D, Tiwari, A (2012) Urban risk assessments: understanding disaster and climate risk in cities. Urban Development. World Bank, Washingon DC. http://hdl.handle.net/10986/12355

Foreword by Anne Golden

Daniel Hoornweg's treatise on Canada's cities comes at a pivotal moment in our history. This book is unique in its scope, looking back and to the future, and in its depth, incorporating a wealth of pertinent data. It will be appreciated by city watchers at every level from academics to policymakers and public officials to practitioners.

Professor Hoornweg is clear-eyed about the challenges and threats, including geopolitical conflict; political polarization thanks to social media which has magnified factionalism and is undermining democracy, relative economic decline; the growing wealth and income gap; climate change and the still growing rate of GHG emissions. Those, like me, who like to have observations about current issues linked to history will find this analysis very satisfying.

The book's focus on urban sprawl as the immediate critical challenge for Canada's cities is noteworthy. I am among those who have been sounding the alarm for decades on urban sprawl as the single most significant obstacle to securing sustainable prosperity for our cities and city-regions. Despite conclusive studies documenting the costs of urban sprawl, despite clear evidence that sprawl is at the heart of productivity-destroying congestion in our major cities, we appear unable to give up our addiction to regional sprawl as the way to manage population growth.

The central message of this comprehensive study is that the solutions can and must be found in Canada's cities. It is cities and metropolitan regions that have the potential to lead the world toward a sustainable, successful future. It is a positive message and could not be more relevant now.

This book was written before the 2024 US presidential election, and, with Donald Trump's return to the White House, the ground has now shifted. What at first seemed like provocative musing by Trump about Canada becoming America's 51st state and the elimination of the "artificial border" separating the two countries has turned into a destructive trade war and a campaign to undermine Canada's sovereignty. This attack has served to underscore the importance of ensuring the survival and success of Canada's major cities. The best way to "Trump-proof" Canada is to ensure that our cities are successful.

Professor Hoornweg's "Halftime Report" for Canada's cities over the 200-year span from 1920 to 2120 (about seven generations) points to reforms needed across many areas. With his expertise on climate change and energy, considerable attention is paid to climate mitigation and adaptation and solutions for environmental degradation.

The agenda for urban reform, reinforced in this study, has been consistent for the past three decades. Solutions must create three enabling conditions: (i) governance, with structures that are big enough to enable region-wide land use and transportation planning, protect the environment, promote economic competitiveness of the region as a whole, and small enough to permit citizen access and local accountability; (ii) municipal finance so that cities have enough money to pay for services and infrastructure that allow them to function; (iii) the huge and growing infrastructure deficit, with comprehensive investment plans for housing, transit and roads, and social and civic infrastructure which anchor its democracy.

The overarching message that resonates most powerfully for me is how essential it is that Canada's cities and city-regions develop a metropolitan mindset that "moves past the traditional zero-sum logic of most politics." As Yuval Noah Harari explains in his first book, *Sapiens*, human beings triumphed over the Neanderthals (with their bigger brains and greater strength) due to our ability to communicate and cooperate. We must stop competing for scarce resources and work together to solve our problems, or we will not succeed.

Cities are where most of Canada's people live and work, where most innovation occurs, where the country's wealth is created, where most immigrants converge, and where people learn to live together with civility. Daniel Hoornweg's "Halftime Report" sets out the challenges with clarity and a compelling sense of urgency. Whether we meet these challenges is up to us.

Toronto, Canada

Anne Golden
Former President and CEO of The
Conference Board of Canada and Chair
of the Greater Toronto Area Task
Force, is a recipient Order of Canada

Foreword by Chibulu Luo

I have had the pleasure of knowing Daniel for nearly a decade, first meeting him when I began my doctoral studies at the University of Toronto in 2015. Even then, his depth of knowledge and passion for cities stood out. I vividly recall him saying, *"When we act in cities, we are not just influencing local dynamics—we are shaping national development and driving global change."* That statement continues to resonate—and it captures the spirit of this important book.

This volume brings together Daniel's extensive body of work on cities, offering insights that span the global to the local—from the dynamics of world cities to the intricacies of Canadian urbanism, with Toronto as a central case. It is essential reading for urban scholars, practitioners, and anyone committed to leveraging the potential of cities to catalyze meaningful transformation.

Focusing on cities is not only strategic—it is foundational. The interconnected nature of urban systems means that cities are deeply embedded in broader ecological, economic, and political processes. Conversations about industrialization, inclusive development, the energy transition, and climate action all converge in cities.

What this book makes clear is that we cannot achieve many of the transitions the world urgently needs—including the "great deceleration" toward sustainability—without rethinking and redesigning how our cities function. I truly hope that Daniel's work will inform and inspire policymakers, planners, and changemakers around the world to invest in urban transformations that advance sustainable development.

Toronto, Canada
Chibulu Luo
Global Energy and Climate Change
Advisor—United Nations
Development Programme (UNDP)

Foreword by Prof. Greg Clark

This book ends with a critical observation. If you were a new human child with the chance to choose where you would be born and live a life that starts today, Canada, and its cities, would be very close to the top of your list. Despite the conscientious, detailed, treatment, and the sober, 'mouth-drying', challenges articulated by our author, a young human today would be very lucky to be born in a Canadian metropolis.

Canada's cities are a worthy focus of attention. The remarkable combinations that make Canada, entwined with the rich social and geographical connections of the Canadian urban footprint and its diaspora populations, coupled with the desire to both cluster together and to reform, make Canadian cities a place of discovery and experimentation in urban futures.

Professor Daniel Hoornweg's halftime report makes for a compelling read. His scholarly approach, dedication to subject, detail, and nuance, provides a rich return for the reader. As a Canadian who has traveled extensively and worked globally, he has the uncanny knack of being able to see Canadian cities from more than one vantage point, at the same time. This provides the reader with a narrator who has more than one lens, and more than one voice. We can read about both the unique strengths, and dire challenges, of urban Canada simultaneously, inviting us to rethink afresh. There is an underlying gestalt to the way this book unfolds, advantages and disadvantages, opportunities and threats. These polarities dramatize the discourses on Toronto, Montreal, Vancouver, Ottawa, Calgary, Edmonton, Winnipeg, Hamilton, Halifax, and their siblings.

Daniel Hoornweg is a special kind of urbanist. He is expert in systems. Those systems are land and infrastructure, water, energy, waste, built environment, and housing, as we would expect. But to this he adds education, health, labor markets, air quality, climate, biodiversity, digital and data, and the wider social, political, and fiscal systems that underpin (or indeed undermine) urban and metropolitan life.

This systems-based approach yields results. It allows us to take an objective perspective and pursue a coherent analysis. Combined with the 200-year viewpoint, this provides us with a clear means to observe the enduring mismatches between policies and finance, governance and aspirations, rights and responsibilities. This is a cogent way to reveal that Canadian cities are not yet on the path of optimization, and serious reform is essential.

A 200-year perspective is a rich lens through which to chart the emergence of urban Canada. One hundred years so far, and one hundred years to go. It is a fertile framework for this half-time assessment. Much achieved. Much still to do. Looming and serious challenges that will not go away and must be addressed.

It is halftime, but it is far from certain whether victory, or defeat, lies ahead. And this is also much more than a halftime score. It is rather the halftime team talk' of the coach at his important urgent work. If Canada's cities are going to win, new tactics are needed.

Fittingly, Daniel Hoornweg dedicates this book to Abha Joshi-Ghani, a mutual friend and colleague, who left us in 2021. Abha would be very proud to have this wonderful book dedicated to her.

Prof. Greg Clark
CBE FAcSS, London

Acknowledgements I am grateful to Ontario Tech University for affording me the time to write this book. The staff at the university library were invaluable, and I thank them.

I am always encouraged by how many people are ready to help build better cities. Some 30 people were interviewed for this book (see Chap. 8). Three people in particular provided help and good cheer: Alan Broadbent, Anne Golden, and Lisa Prime. Greg Clark, in London, showed me how business improvement areas (BIAs) got their start in Toronto, and how, like many good ideas from cities, the concept was widely distributed around the world.

I try not to take advantage of having one of the world's best in-house editors; however, my wife Jacquie makes that difficult as she is not only an excellent editor, but also has a depth of knowledge and interest in cities and sustainability. Her help with this book is truly appreciated, and I am fortunate to have Jacquie's encouragement, advice, and love.

Last, but never least, are the professionals and citizens who toil every day to design, build, and manage our cities. The mayors, councilors, planners, engineers, receptionists, reporters, waste collectors, street sweepers, snowplow operators, accountants, and the many people who run the world's cities help us all. Cities are humanity's greatest achievement, and because their impacts ripple across all ecosystems, there is no work more important than the work of cities. Thank you all.

Competing Interests The author has no competing interests to declare that are relevant to the content of this manuscript.

Praise for *Canada's Cities in a Changing World 1920–2120*

"This book is a fantastic read. It provides deep insights on the physical, economic, environmental, social, and historical aspects of Canada's cities with a very modern, clever, and relatable perspective by the author. The book is grounded in rich examples of the opportunities and challenges for Canada, with important takeaways for the future in a complex and ever-evolving urban world."

—Judy Baker, *Senior Advisor, Sustainable and Inclusive Cities*

"In *The Halftime Report*, Daniel Hoornweg draws on his extensive practical experience working with cities worldwide to examine urban history and project urban futures for Canadian cities. The book is filled with valuable insights for cities globally and offers a much-needed longitudinal perspective during this volatile moment in time."

—Raf Tuts, *Director, Global Solutions Division, UN-Habitat*

"From Adam Beck to David Foot, Jevon's paradox to the Johari window, Daniel Hoornweg has produced an invaluable volume for Canadian city watchers, a kind of compendium to understand where we've come from and potentially headed toward. There's lots to be inspired—and shamed—by here, and he's created a terrifically credible analysis of what's at stake for its cities and Canada."

—Mary Rowe, *President and CEO at the Canadian Urban Institute*

"Daniel Hoornweg brings such a richness of detailed expertise and broad experience with cities to provide a many-layered history and understanding of Canada as an urban nation. As a civil engineer, planner, risk manager, political scientist, and a resident and sociological observer, the book journeys the evolution of Canada's cities in a changing world system, zooming in and out between fascinating nitty-gritty and global complexity. The route he has crafted truly has me re-learning my home city (Toronto) and country again."

—Jeb Brugmann, *Founding Secretary General, ICLEI-Local Governments for Sustainability; Author, Welcome to the Urban Revolution: How Cities Are Changing the World*

"*Halfway There* by Daniel Hoornweg is an essential read for policymakers, planners, and leaders in rapidly urbanizing countries. Drawing from Canada's experiences, both successes and shortcomings, the book offers a deeply insightful roadmap for managing the enormous challenges that cities worldwide are now facing. With its emphasis on resilience, sustainable urban growth, metropolitan governance, and systems thinking, the book transcends national boundaries. It provides emerging economies with practical frameworks to anticipate and manage explosive urbanization, climate risks, migration pressures, and economic transitions. Hoornweg's vision empowers cities to act as engines of sustainability and innovation, making this work highly relevant for African, South Asian, and other developing regions poised to shape the 21st-century urban future."

—Ming Zhang, *Global Director, Urban, Resilience, and Land Global Department, The World Bank*

Contents

1	**Introduction**	1
	References	4
2	**Cities in a Changing World**	7
	2.1 The Kind of Problem [and Opportunity] a City Is	12
	2.2 The Metropolitan Voice	15
	2.3 The Century of Countries	16
	2.4 An Uneven Relationship	17
	2.5 The Power of Geography	18
	2.6 The City as a Wicked Problem	19
	2.7 Intervening in Urban Systems	20
	2.8 The City as Commons	22
	2.9 Canada, Growing Trouble	24
	2.10 The Measure of Cities (Census Metropolitan Areas and Agglomerations)	26
	2.11 The Impact of Humans: A Geological Perspective	27
	2.12 The Shift to Sustainability	29
	2.13 The Power of Hierarchies—A Fable	30
	2.14 Networked Hierarchies and the Superpower of Cities	31
	2.15 Green from the Grassroots: Elinor Ostrom Championing the Power of Cities	31
	References	32
3	**Jazz in the Kitchen: The Special Case of Toronto**	35
	3.1 Why so Many Cooks?	37
	3.2 The Limits of Boundaries	40
	3.3 Transportation in the Toronto Region—A Spoiled Broth	42
	3.4 The Taylor Swift Eras Tour—A Sour Note for Toronto	45
	3.5 Trends Affecting Toronto	45
	3.6 The World Needs More Toronto	49
	3.7 Toronto Region to Toronto Global?	52
	References	54

4	**Booms, Busts, and Echoes Around the World: Demographics and the Great Acceleration**	**57**
	4.1 Boom, Bust, and Echo: Canada's Continued Contribution?	63
	4.2 From Immigration to Migration	64
	4.3 The Rise of Post-secondary Education	66
	4.4 Booms and Busts in Canadian Cities	66
	4.5 Canada's Population in 2120	69
	References	71
5	**The Wealth and Waste of Cities**	**75**
	5.1 The Waste and Wealth of Canada's Cities	78
	5.2 What a Waste	82
	5.3 A Visitor's Perspective	84
	5.4 Cities and the Jevons Paradox	84
	5.5 The Values of Gold: Resource Demands of Canadian Cities	86
	5.6 Resources Canada	87
	5.7 Leave It on or in the Ground: Beyond Beavers, Bison, and Oil	93
	5.8 Blame Canada, Rich Canadians, or All the Rich?	95
	References	100
6	**Seven Generations of Canada's Cities: The Halftime Report**	**105**
	6.1 Demographics and Productivity	111
	6.2 Climate Change	112
	6.3 Energy and Material Flows	115
	6.4 COVID-19	116
	6.5 State of Not-so-Good Repair	117
	6.6 London Burns	117
	6.7 Follow the Trends	119
	6.8 One Number to Rule Them All: Looking Beyond GDP	119
	6.9 Looking Beyond GDP	121
	6.10 The Difficulty in Measuring the 'Ease of Doing Business'	122
	6.11 The Next Summit	123
	References	125
7	**Brave the Future: The Road Ahead**	**129**
	7.1 The Tenacity of Cities	133
	7.2 Pebbles in the Pond	134
	7.3 Competition and Cooperation	135
	7.4 The Roots of Immigration—The Need for Querencia	137
	7.5 From Scarcity to Sufficiency	140
	7.6 Fracking Our Cities	141
	7.7 Tear Down These Walls	143
	7.8 From Metropolitan to Sustainability Mindset	145
	7.9 The Twenty-Ninth Day	147

7.10	Canada's New Narratives	147
7.11	Braving the Futures—Diverging Futures for Cities and Countries?	150
7.12	Out with a Boom?	152
7.13	Municipal Pooling	153
7.14	Recommendations	155
7.15	Council's Climate Resolution	158
	References	163

8 Thoughts of Others ... 167
   8.1 Challenges ... 169
   8.2 Opportunities ... 171

9 Conclusion, A Call to Action ... 173
   References ... 183

List of Tables

Table 2.1	World's growing cities and wealth	8
Table 2.2	Regional share of the world's largest 100 cities, 1800–2100 (%)	10
Table 2.3	World's largest cities (urban areas) by population (millions), 1500–2100 with the largest Canadian Comparator	11
Table 3.1	Population of the greater golden horseshoe (Toronto region)	38
Table 4.1	Canada population 1921 to 2120	67
Table 4.2	Changes to populations July 1, 2022, to June 30, 2023	68
Table 4.3	Population of Canada's largest cities 1871–2021	71
Table 5.1	Distribution of Canada's GDP in 2020 and estimated 2070	78
Table 5.2	Primary energy use per person in 2022	79
Table 5.3	Per capita GDP in 'Western Offshoots', 1600–2020 (1990 dollars)	80
Table 5.4	Globalization and World Cities (GaWC) City Ranking [75]	85
Table 6.1	Global events and trends	110
Table 6.2	Top 10 disruptions by likelihood and impact	112
Table 6.3	Trends in urban Canada	113

List of Boxes

Box 3.1	Rolling Up the Sidewalk: Toronto's Response to Google	52
Box 5.1	Scope of the Challenge, a Consumer's Tale	97
Box 5.2	A Measure of Success; US Embassies and Local Air Quality Around the World	99
Box 5.3	The Second Waste Hierarchy. Politics Matters Most	99

Chapter 1
Introduction

Before their 80th birthday, a child born in 1920 would have seen the world's population triple, from less than 2 billion to more than 6 billion, while global wealth, and waste, increased more than 40-fold. This massive increase in wealth and waste picked up speed as the *Great Acceleration* started in earnest around 1950. Canada's cities contributed more to this overall increase in wealth and waste than cities in any other country.

Today's children will likely see during their lifetime the cresting of two massive waves. World population is on track to peak in the next 50-to-60 years. The second peak, connected to the first, is more difficult to discern, but is out there and will hopefully arrive alongside peak population.

Peak waste, or peak pollution and planetary impact, is the point where humanity's aggregate impact on the planet's biosystems crests and then also starts to decline. Hopefully the decline is steep, even though some impacts, like carbon dioxide emissions, will take hundreds of years to ameliorate, and impacts like biodiversity loss are permanent. The arrival of peak waste (greenhouse gas emissions, solid waste, wastewater, ecosystem degradation) needs to be led by those who are having the biggest impact today. No community is better placed to lead this *great deceleration* and massive shift toward sustainability than Canada's cities.

The exact date and magnitude of peak population are not yet known, but every year it becomes a bit clearer. The rate at which population declines after this peak is also unknown but has major consequences. Before a semblance of sustainability emerges, today's children will likely see continued massive changes to planetary systems such as another 30 percent loss of biodiversity and global average temperatures increasing above 2.5 °C. These changes will bring about geopolitical shifts and levels of migration so severe that the fabric of today's nations will likely be torn and will require repair as a new tapestry of humanity emerges.

If that child spends much of their life in Canada, they will be more fortunate than most. Much can be done today to increase the likelihood of this good fortune.

The best place to intervene to bring about this better fortune is in cities, especially Canadian cities.

Most of Canada's cities were communities first established by Indigenous peoples. The confluence of rivers, the terminus of a portage, proximity to fertile fields or productive waters, Canada's cities grew from the land. Most of Canada's cities—for example, Montreal, Toronto, Winnipeg, and Vancouver—were established long before Canada started as a country. And like cities around the world, they built their country, and they will likely outlast the structure of today's countries. Cities offer permanence: a place to shelter in the storm.

The 1920s are a useful place to pick-up the story of Canada's cities. Canada's urban shift began in the 1920s as cities became home to more than half the country. The decade was turbulent but mostly prosperous for Canada. The country was recovering from the hard-fought WWI, which affirmed Canada's place as an independent nation. Canada insisted on a separate seat at the newly created League of Nations. In 1923, Canada signed its first international agreement, the Halibut Treaty with the USA, without Britain's involvement.

Cinemas, automobiles, radios, airplanes, bootleg booze, and jazz clubs, especially in Montreal, seemed to usher in a modern age. The Roaring Twenties may have started slowly with almost 15 percent unemployment from returning soldiers and prairie farmers suffering from low wheat prices (in 1921 one-third of Canada's jobs were in the agriculture sector). But by the mid-twenties, things were markedly improving. Demand from the USA for wheat and timber, metals, and increasingly manufactured goods was rising. Canada came into its own in the 1920s. Women received suffrage in 1918 (although Asian women and men were not granted suffrage until 1948).

Decreases in infant mortality and increases in overall life expectancy illustrate well how Canada began to flourish in the 1920s. About 100 in 1000 (i.e., 10 percent) Canadian babies died within the first year of life in 1921 (with rates as high as 136 deaths per 1000 live births in Toronto [1]), compared with about 4.4 per 1000 live births in 2021 [2]. Similarly, the life expectancy at birth for men increased from 58.8 years in 1920–1922 to 79.3 years in 2009–2011. During the same period, the life expectancy of women increased from 60.6 years to 83.6 years.

Amazingly, the 22 years of increased life expectancy experienced in Canada over the last 100 years is still less than the differences today between some neighborhoods in the same city. For example, there is a 23-year gap in life expectancy between the highest life expectancy neighborhood in Hamilton (87.7 years) and the lowest (64.8 years) [3].

A halftime report for a 200-year time horizon, about seven generations, provides a timely opportunity to see how large-scale infrastructure, social norms, and geopolitics are influencing cities. Governing and managing over Seven Generations, as introduced by the Haudenosaunee Five Nations Confederacy (later six nations), helps temper hubris, and hopefully provides sufficient perspective to identify ways to work toward sustainability, and flourishing for all.

Compared to cities in other high-income countries, Canada's cities are unique in that they contribute a disproportionately large share of the national economy. Toronto,

Canada's largest city contributes 27 percent of Canada's economy, while New York City the largest US city contributes just five percent of the country's economy.

Canada's cities are among the world's most energy- and material-intensive. Vancouver, Calgary, Montreal, and Toronto are often recognized as some of the world's most livable cities. However, if the rest of the world lived like Canadians, the planet's biosystems would be even more overwhelmed. Per person, a Canadian causes about 10 times more environmental degradation than the average global resident, and more than 30 times someone living in a low-income country.

Politically, Canada's larger cities are often at odds with broader national sentiment as well as their own provincial governments. Canadian cities typically have fewer fiscal levers and management tools than their international comparators. For example, of all taxes raised in Canada, less than 10 percent contributes to municipal budgets [4], even though municipalities pay for about half of all urban infrastructure.

A city is always a work in progress, a complex *emerging* system. Cities are much longer-lived than the businesses they support, as well as countries that do not enjoy the same scaling powers of their cities. A large city serves as a frontier zone for disparate people with diverse backgrounds who come together and develop new ideas, new ways to generate wealth, and advances in culture. The city serves as the port for importing and exporting these ideas. A good city is a great place to learn.

Canadian cities, especially the larger cities of Vancouver, Calgary, Edmonton, Toronto, Ottawa-Gatineau, and Montreal, need enhanced mechanisms to foster collaboration. Clear metrics to encourage higher productivity and enhanced sustainability are available and should be published annually by Statistics Canada. The federal government and possibly the provinces should allocate a share of the harmonized sales tax (HST) directly to local municipalities as a function of local GDP growth and sustainability targets. Urban areas should be able to raise revenues through their streets, e.g., charging by vehicle kilometers traveled and local data systems. However, this needs to be predicated on enhanced collaboration across local governments, and most important, increased trust. A city has agency, but this agency is best exercised by the collective.

Canada is buffeted by several growing threats. These include political polarization; geopolitical conflict; degrading public communications and disinformation; relative economic decline; a growing wealth gap; significant shifts in happiness, or well-being, between age cohorts (largest gap in the OECD); and an uncertain desire and ability to safeguard earth systems, e.g., limited support to reduce greenhouse gas (GHG) emissions.

Canada's cities will need to lead in both climate mitigation and adaptation. They will also need to lead in the challenges of accommodating growing numbers of migrants.

From the time of first European colonization until the mid-nineteenth century, the native population of Canada dropped from roughly 500,000 (some estimates put it as high as two million) to just over 100,000. This was due to a mixture of disease, starvation, and warfare, instigated by European migration to the region. The native population was generally segregated and oppressed into the second half of the 1900s.

Canada's last residential school, in Punnichy, SK, closed in 1996. Native Canadians were only given the vote in 1960.

Despite the difficult history, the Canadian government has made progress in trying to include Indigenous cultures in the country's national identity. As of 2020, Indigenous Canadians make up more than five percent of the total Canadian population. About half of those live in cities. Indigenous peoples, the first residents of most of Canada's cities, will need to help shape how Canada's communities interact with the land, and the world. How Canada receives and integrates migrants is a key aspect of reconciliation with First Nations.

Canada's economy boomed in the aftermath of the WWII, and a stream of social programs such as universal health care and pension plans were introduced. These contributed to a rise in the standard of living. However, Canada's impact on the environment grew in lockstep with the burgeoning economy.

Trends and events over the next 100 years will threaten Canada's prosperity. Canada's cities will need to buttress themselves for these impacts, manage through the transition, and support the country as it navigates the currents of circumstance. Two new narratives are proposed to move cities toward a sustainability mindset, namely (i) migration within sustainable development and (ii) the shift from a scarcity to sufficiency paradigm.

We build our cities, and our cities build us and our societies. To do this, cities require two types of freedom. Freedom *from* and freedom *to*. Freedom from strife, crime, disease, and disaster is prioritized by provincial and national governments. As many of the most challenging issues and opportunities for the next 100 years are collective and cross borders, cities also need freedom *to* work innovatively and cooperatively. Cities need to work with citizens to better define progress, growth, and sustainability.

The last 100 years in Canada could be summarized as the age of plenty. The economy grew by a robust 2.5–3.0 percent per year. Canada's relative global influence peaked around 1975. The post-WWII years were mostly peaceful. The closest Canada came to war was shooting at a Spanish fishing boat. The next 100 years may be very different. Canada and its cities are shifting to an age of extremes. Climate extremes, political polarization, growing wealth and income disparity, and probably many more discussions on migration—numbers, origin, and ways to assimilate newcomers.

References

1. Hallman S (2009) The effect of pandemic influenza on infant mortality in Toronto, Ontario, 1917–1921. In: A thesis submitted for master of arts. McMaster University
2. Statistics Canada. https://www150.statcan.gc.ca/n1/pub/11-630-x/11-630-x2016002-eng.htm
3. McMaster University (2019) Starting the conversation about social inequality and healthy aging
4. Kitchen H, Slack E (2016) More tax sources for Canada's largest cities: why, what, and how? In: IMFG papers on municipal finance and governance

Open Access This chapter is licensed under the terms of the Creative Commons Attribution-NonCommercial-NoDerivatives 4.0 International License (http://creativecommons.org/licenses/by-nc-nd/4.0/), which permits any noncommercial use, sharing, distribution and reproduction in any medium or format, as long as you give appropriate credit to the original author(s) and the source, provide a link to the Creative Commons license and indicate if you modified the licensed material. You do not have permission under this license to share adapted material derived from this chapter or parts of it.

The images or other third party material in this chapter are included in the chapter's Creative Commons license, unless indicated otherwise in a credit line to the material. If material is not included in the chapter's Creative Commons license and your intended use is not permitted by statutory regulation or exceeds the permitted use, you will need to obtain permission directly from the copyright holder.

Chapter 2
Cities in a Changing World

Since the dawn of time, empires have had one thing in common; their center of power has been led by the metropolises within them.

Continents and empires have been run by a few metropolises. St. Petersburg and Moscow, Beijing and Nanjing, Vijayanagar and Delhi, Mexico City, Machu Picchu, Washington DC, Venice and Rome, Amsterdam, Madrid, Istanbul, Zanzibar, Zimbabwe, Lagos and Kumasi; these cities served as nodes for power, anchoring vast networks. Tables 2.1, 2.2 and 2.3 illustrate how the world's cities have ebbed and flowed with centers of power shifting first from Asia with cities like Beijing to Europe and London and Paris. From there to New York and Los Angeles in America, the focus is now shifting back to Asia and cities like Shanghai, Mumbai, and Jakarta. The final shift will be over the next 50-to-75 years with the rise of cities in Africa like Lagos, Dar es Salaam, and Kinshasa.

From the first civilizations that relied on agriculture and permanence of residence, it took the world almost 10,000 years to reach a population of a billion people. From one billion in 1820, it took 123 years for the world to reach two billion people in

Table 2.1 World's growing cities and wealth

Year	GDP (per person)	Population (Mn)		Number of cities (Canadian) (1–5 Mn, 5–10 Mn, > 10 Mn)
		Global	Urban	
1500	$130	461	< 40	1, 0, 0
1600	150	554	< 50	1, 0, 0
1700	170	603	< 60	1, 0, 0
1800	200	990	< 100	1, 0, 0
1900	680	1650	320	11, 1, 0
1950	3050	2487	746	40 (2), 4, 1
2000	6500	6144	2950	350 (3), 31 (1), 19
2050	16,500	9602	6390	680 (4), 73 (1), 50 (1)
2100	40,000	10,853	8686	800 (5), 91 (1), 64 (1)

Adapted from Hoornweg and Pope, in Labbé, D., & Sorensen, A. (Eds.). (2020)

1927. From there, growth gathered speed. It took 33 years to reach three billion in 1960, and just 14 years to reach four billion in 1974. Even more quickly, in 13 years global population reached five billion in 1987, and 12 years to reach six billion in 1999 and seven billion in 2011. In only 11 years, population grew to 8 billion in 2022. The latest predictions from here show population growth slowing, with the world reaching nine billion in 2037 (taking 15 years), and then growth further slows, hitting a peak population of 10.8 billion around 2086.

Even more important, from a global ecosystem and economic perspective, is the rate of global urbanization. In 2008, the world surpassed 50% urban. Just a few years before that, Canada reached 80% urban in 2006. This is likely the 'peak urban' level that the rest of the world will also reach early in the twenty-second century. Every day the world's urban population swells by 200,000 people. These massive population and urbanization trends are now meeting increasingly strong headwinds of earth system responses like climate breakdown, biodiversity loss, and the too slow transition to sustainability.

For the last 100 years, what countries were doing, often crowded out what cities were doing. More than 120 new countries emerged in the last century. Political debate was often about a country's place in the world and shifting power and wealth internationally. This "century of countries" is now being followed by the "century of cities" [1].

Countries will continue to project power, attempt to safeguard access to energy and material flows, and limit the movement of people and information. However, cities, the agglomerations of people and economy, will increasingly be tasked with their own resilience and the day-to-day implementation of sustainability. For the next 100 years the degree of sustainable development will mostly be decided by the pragmatism and on-the-ground actions in cities.

In 1900, the global population was 1.7 billion, with only 320 million people living in cities. Urban populations increased some ninefold over the last 100 years and are on track to double again and increase to more than nine billion people by 2120. As cities grow, they buttress the world's countries. Cities existed before most countries and will likely be around long after their host countries have disappeared. Cities largely gave finance and legitimacy to countries. Cities grew the world's economy and drove the energy and material flows that underpinned the *Great Acceleration.* Trade, for example, is mostly heading to customers in cities.

Cities (urban areas) are the most critical unit of analysis for sustainable development. Fortunately, a science of cities emerged over the last 50 years, typified through research by Bettencourt [2], West [3] and Batty [4]. Urban areas mimic natural systems and therefore benefit from a systems approach as outlined by people like Donella Meadows [5]. The *city as commons,* as discussed by Ostrom [6] is also emerging as an optimistic view, giving pause, and some hope against Hardin's *Tragedy of the Commons* [7].

Hardin's tragedy of the commons suggested that resources like a common pasture would be overwhelmed with too many farmers releasing too many sheep. Ostrom's optimism highlighted several examples, where if local people were given the tools, they could effectively manage common resources like fish stocks and forests.

Canadian cities were largely shaped by geography and a resource development mentality. This mindset however is no longer fit-for-purpose as the world approaches peak population; the locus of economy shifts from OECD-member countries to Asia and then to Africa, and society deals with climate change and other earth system transformations. A few tools for improved urban management include application of hierarchies, strategically intervening in urban systems, urban metrics, biomimicry, and nature-based solutions. Urban managers can also benefit from a broad overview such as a geological perspective, and an appreciation of the permanence and vulnerability of cities. This is especially relevant as the world moves into a century of much greater wealth, urban populations, significant climate change, and massive geopolitical and technological change.

The Hanseatic League of Cities operated from the late twelfth century to the mid-seventeenth century. At its height in the fourteenth century, the League encompassed about 200 cities across seven modern-day countries [8]. The League ended when the 1648 Treaty of Westphalia largely replaced city partnerships with nation-states. More recently, the global pattern of nation-states fractured extensively. From 1900 to 2000, the number of countries in the world quadrupled (~ 196 today, plus territories).

The latter half of the twentieth century saw the Great Acceleration where wealth and material flows grew enormously (anchored in cities). Cities have also grown enormously in size. From 1700 to 1800, there was one city with a population above one million (Beijing). In 1900, only London was larger than one million (Beijing had dropped to 725,000) [9], yet by 2000 there were more than 350 cities with populations above one million: a number likely to swell to 1000 cities by 2100 (Table 2.1). Energy and material flows for this massive urban expansion are staggering. So too the corresponding environmental impacts [10].

In 1950, New York was the world's only megacity with a population over 10 million. Today there are about 45 megacities, and by the end of the century, there will likely be 64 megacities (including Toronto).

Table 2.2 Regional share of the world's largest 100 cities, 1800–2100 (%)

Region	1800	1900	1950	2000	2050	2100
World	100	100	100	100	100	100
Africa	4	2	3	8	21	36
Asia	65	22	37	44	48	48
Europe	28	53	34	15	6	3
Latin American and Caribbean	3	5	8	16	13	6
Northern America (**Canada**)	0	16 (**1**)	16 (**2**)	15 (**2**)	11 (**1**)	7 (**1?**)
Oceania	0	2	2	2	1	0
Average size of world's 100 largest cities	187,000	725,000	2.2Mn	6.3Mn	13.9Mn	19.2Mn

Adapted from Hoornweg and Pope, in Labbé, D., & Sorensen, A. (Eds.). (2020)

In addition to cities growing in population, the location of the world's larger cities is changing (Table 2.2). In 1800, 65 of the world's 100 largest cities were in Asia. By 1900, the locus of urban (and global) power had shifted; 53 of the world's largest cities were in Europe. However, from 1900 to 2100 the number of larger cities in Europe is expected to plummet from 53 to three, while Africa increases from two to 36, and Asia, which declined to 22 in 1900 (from 65) increases to 44. Canada had two cities, Montreal and Toronto, in the world's Top 100 for about 50 years, however Montreal dropped out of the list in the 1970s. For the rest of this century, Toronto will be the only Canadian city in the Top 100 list.

Table 2.3 provides an estimate of the population of the world's largest cities from 1500 to 2100. In 1500, Beijing, with a population of about 670,000, was the world's largest city. In 2100, Lagos is projected to be the world's largest city with a population of almost 80 million. Table 2.3 provides population estimates based on urbanization trends. When viewing the projected ten largest cities in 2100, ranging in population from 40 to 80 million, the potential impacts of climate change become particularly worrisome. The ten largest cities—Lagos, Dar es Salaam, Kinshasa, Mumbai, Karachi, Delhi, Kolkata, Luanda, Dhaka, and Nairobi—are all in Africa and Asia. These geographic regions are expected to be especially impacted by a warming world [11]. Kinshasa and Dhaka already have seen major out-migration from climate-related heat stress, flooding, and related conflict.

Cities are a mix of residents and visitors, along with pets and wildlife, buildings and infrastructure, vegetation, businesses and institutions, all reflecting the local geomorphology and climate. Vancouver, situated between mountains and the sea, overtop an active tectonic zone, next to the Pacific Ocean, with its wet temperate climate, evolved differently from Montreal. Montreal, with its robust winters, anchored to sedimentary rock atop the Canadian Shield and further upstream on the St Lawrence River than Quebec City, served to connect much of Canada to Europe. Montreal was Canada's largest city for about 120 years.

Table 2.3 World's largest cities (urban areas) by population (millions), 1500–2100 with the largest Canadian Comparator

1500	1600	1700	1800
Beijing (0.67)	Beijing (0.71)	Istanbul (0.7)	Beijing (1.1)
Vijayanagar (0.5)	Istanbul (0.7)	Tokyo (0.69)	London (0.86)
Cairo (0.4)	Kyoto (0.3)	Beijing (0.65)	Guangzhou (0.8)
Hangzhou (0.25)	Hangzhou (0.27)	London (0.55)	Tokyo (0.69)
Tabriz (0.25)	Paris (0.24)	Paris (0.53)	Istanbul (0.57)
Gauda (0.2)	Naples (0.22)	Ahmedabad (0.39)	Paris (0.55)
Istanbul (0.2)	Cairo (0.2)	Esfahan (0.35)	Naples (0.43)
Paris (0.19)	Bijapur (0.2)	Kyoto (0.35)	Hangzhou (0.39)
Guangzhou (0.15)	Nanjing (0.19)	Hangzhou (0.3)	Kyoto (0.4)
Nanjing (0.15)	Ahmedabad (0.19)	Bijapur (0.26)	Moscow (0.25)
Various settlements	**Various settlements**	**Quebec City (0.008)**	**Montreal (0.02)**
1900	1950	2000	2050	2100
London (6.5)	New York (12.4)	Tokyo (34.5)	Mumbai (47.4)	Lagos (79.8)
New York (4.2)	London (8.9)	Mexico City (18.1)	Delhi (40.2)	Dar es Salaam (62.3)
Paris (3.3)	Tokyo (7.0)	New York (17.8)	Dhaka (37.5)	Kinshasa (60.3)
Berlin (2.7)	Paris (5.9)	São Paulo (17.1)	Karachi (36.9)	Mumbai (57.7)
Chicago (1.7)	Shanghai (5.4)	Mumbai (16.1)	Kolkata (36.8)	Karachi (49.9)
Vienna (1.7)	Moscow (5.1)	Kolkata (13.1)	Lagos (36.3)	Delhi (48.9)
Tokyo (1.5)	Buenos Aires (5.0)	Shanghai (12.9)	Tokyo (35.1)	Kolkata (44.7)
St. Petersburg (1.4)	Chicago (4.9)	Buenos Aires (12.6)	Kinshasa (33.3)	Luanda (42.3)
Manchester (1.4)	Ruhr (4.9)	Delhi (12.4)	Mexico City (29.8)	Dhaka (42.3)
Philadelphia (1.4)	Kolkata (4.8)	Los Angeles (11.8)	Cairo (27.9)	Nairobi (38.4)
Montreal (0.36)	**Montreal (1.34)**	**Toronto (4.7/ 5.9)**	**Toronto (13.5/ 18.1)**	**Toronto (21/ 27.4)**

From: From Hoornweg and Pope, in Labbé, D., & Sorensen, A. (Eds.). (2020). Toronto population for GTHA/Toronto Region. In 1600 approximately 250,000 First Nations and Inuit lived in what is today Canada. Port Royal was established as Canada's first permanent settlement in 1605, Quebec City in 1608

Cities are also dense areas of economic activity made possible through the stock and flow of energy and materials, as well as ideas and entrepreneurialism. Building materials are typically sourced from as near-by as possible; however, larger cities trade ever more-distant foodstuffs, materials, and increasingly services and finance. Montreal, for example, anchored much of the continental trade in beaver pelts for almost 250 years, in some years shipping 100,000 pelts for European hat-making.

In Canada, urban areas drive more than 80% of the country's GDP, energy demand, material use, and solid waste generation, along with overall GHG emissions. The care and feeding of a city is an enormous task, especially in Canada where cities are among the most energy and materials hungry in the world (Chap. 5).

Cities, migration, and trade were the main catalysts of progress in the developed world over the past two centuries [12]. Canada and its cities developed with a relatively permeable national border between Canada and the USA, and relatively impermeable borders between provinces. Alvarez et al. [13] among others, reviewed internal trade barriers in Canada and make a strong case for liberalization. Removing non-geographic trade barriers within Canada could provide an initial 3.8% increase to the country's GDP [14]. The utilities, education and health, telecommunications, and business services sectors are most impacted, with costs of trade barriers exceeding 50%. The equivalent internal trade barriers are largely non-existent in the USA. A similar drain on economic productivity is attributed to Canada's agricultural marketing boards (raising costs by more than $39 billion per year) [15]. The implicit tax rate of supply management (as share of income) is much higher for low-income households than for high-income households. In addition, Canada's dairy supply management system has led to over 6 million liters of milk wasted since 2012 (with a financial loss of $14.9 billion and 8.4 million tons of CO_2 emitted) [16].

Agricultural supply management and interprovincial trade barriers were ostensibly established to support regional and provincial demands. The costs attributed to these policy interventions are however disproportionately borne by customers in urban areas (as this is where more than 80% of Canada's economy is derived, and most agricultural products are sold). This is like national, and provincial, debt which makes a claim on future productivity, which again will be disproportionately borne by urban economies in Canada (see Chap. 6).

In addition to bearing much of the economic impact associated with the complexities of federal-provincial relations, Canadian cities also encouraged low energy prices, auto-based residential land development, and relatively low-cost land. Compared to European cities, Canada's cities use about twice as much electricity per household (and electricity costs are about half the household cost than those in Europe). Canada's cities being some of the least dense in the world, also have relatively high infrastructure costs (more roads and infrastructure per person). This also contributes to high per capita greenhouse gas emissions (see Chap. 5).

2.1 The Kind of Problem [and Opportunity] a City Is

In the last chapter of *The Death and Life of Great American Cities*, Jane Jacobs provides a framework to answer the question on *The kind of a problem a city is* by defining cities as problems in *organized complexity* [17]. This approach is based on her association with Warren Weaver and an appreciation of his 1948 paper *Science and Complexity* [18].

2.1 The Kind of Problem [and Opportunity] a City Is

"What makes an evening primrose open when it does? Why does saltwater fail to satisfy thirst?" "Why does the amount of manganese in the diet affect the maternal instinct of an animal?" Weaver explored a class of problems which deal simultaneously with a sizable number of factors that are interrelated into an organic whole. They are not *problems of disorganized complexity*, to which probability and statistical methods hold the key. Rather, he argued they are problems of *organized complexity*. Warren also opined, "A revolutionary advance must be made in our understanding of economic and political factors. Willingness to sacrifice selfish short-term interests, either personal or national, in order to bring about long-term improvement for all must be developed" [18].

More than 50 years later, Luis Bettencourt returns to the challenge raised by Jane Jacobs on a city's organized complexity in his essay *The kind of problem the city is: New perspectives on the nature of cities from complex systems theory* [19]. Bettencourt illustrates how clearer formalizations of social networks and conceptualization of cities as complex adaptive systems provide strong support for urban policy and principles.

Cities (urban areas) behave like natural systems. The patterns of city commuters have an uncanny resemblance to tributaries, streams, and rivers, or capillaries and blood vessels in a body, or migrating caribou. Increasingly, the effective boundaries of urban areas can be defined more organically through data such as that from cell phone coverage [20].

Through the flow of energy and materials, scaling laws emerge in inanimate systems (geophysical), the same way they emerge in animate systems (biological or allometric) [21, 22]. Urban life follows a similar pattern of hierarchies and strengthening flows of people, energy, traffic, and waste. This is why people in larger cities, for example, tend to walk faster than those in smaller communities [23].

A powerful attribute of urban systems is that they follow the power law (Eq. 2.1). A doubling of city population can lead to more than a doubling of social interactions and economic output with less than twice the demand for new infrastructure. Larger cities typically provide more wealth for less (infrastructure) cost [24].

$$\Upsilon(t) = \Upsilon_0 N(t)^\beta \tag{2.1}$$

In Eq. 2.1, Y represents the urban indicator, e.g., GDP or meters of electrical wiring, while N represents the city population at time (t). Y_0 is a time-dependent constant. β is the scaling component. Considerable empirical evidence illustrates that when city populations double β is ~ 1.15 (superlinear) for social interaction outcomes like economy and patents; and β is ~ 0.85 (sublinear) for infrastructure such as road surfaces and length of transmission piping [25].

Scaling laws for cities suggest a resident in Saguenay may well expect Montreal residents to contribute more per person to Quebec's and Canada's economy, or North Bay residents with those in Toronto, Slave Lake and Edmonton, or Prince George residents and those in Vancouver. Similarly, residents of Canada's larger cities can expect that almost all infrastructure costs are lower per person than those in more remote, smaller communities, and that these differences are growing. What residents in both larger and smaller communities however need to appreciate is the changing role of urban economies (wealth generation and infrastructure provision) across Canada, how this changed over the last 100 years, and how this change will be even more significant over the next 100 years as global populations shift markedly.

A large urban area naturally wants to follow scaling laws; however, a border can act the same way a fence might interfere with the movement of animals across the savannah. Staunch the flow of people, ideas, finance, or materials, and system inefficiencies increase. Trade barriers between provinces are an example of these imposed inefficiencies.

As cities grow so do externalities like congestion, pollution and crime. These also threaten efficiencies and urban well-being. However, policy interventions to address these should be implemented in a way that maintains the benefits of larger cities and the benefits of scaling. For example, encouraging congestion pricing and basic minimum incomes.

In addition to benefitting from the scaling aspects of larger cities, an urban systems approach is strengthened through the application of tools such as urban metabolism, hierarchies, and biomimicry, i.e., borrowing concepts from nature to build and manage cities.

As a mix of people, energy and materials, cities drive culture and economy. National economies reflect urban hierarchies where one or two key cities are at the apex of the hierarchy and generate a disproportionate share of the economy (as well as direct and indirect environmental impacts). These larger cities will continue to connect globally and reinforce inherent international hierarchies. Canada's three largest cities, for example, are home to more than a third of Canadians, and almost half the country's economy (both shares are growing).

The opportunity of cities arises from their ability to bring things together. Cities concentrate people in space and time and enable more intense use of infrastructure. Agglomeration benefits include higher productivity, lower transportation costs (suppliers can be closer), increasing markets and employment opportunities, more opportunity to specialize, and knowledge spillovers between researchers and firms, sparking new ideas and innovation.

The collapse of complex societies, such as cities, is usually experienced as the 'loss of complexity' [26], for example, squabbles of the elites or recalcitrance of community factions. In system dynamics, change mostly happens at the emergent edges. Tidal flats, marshes and tropical rainforest support increased biodiversity as 'edge cities', suburbs and diverse neighborhoods, support cultural adaptations and business incubation.

As complex adaptive systems, cities have an evolutionary drive. This, in addition to Canada's cities being conditioned to more than 100 years of 2-to-3% annual economic growth (and population increase), makes declining populations and slowing economies especially difficult to deal with. Like giant sloths, mammoths, and elephants, cities will need to adapt and evolve into smaller, less energy and material consumptive, versions to avoid collapse. Of the higher-income OECD-member countries, Canada's larger cities are likely to be among the last that need to deal with declining populations; however, every Canadian city should be anticipating declining populations within 100 years (possibly 50), and dramatically declining energy and material consumption as soon as possible.

2.2 The Metropolitan Voice

> I am the Lorax! I speak for the trees. I speak for the trees, for the trees have no tongues.
> — Dr. Seuss

Metropolitan areas such as Calgary-Edmonton and Greater Toronto are important urban systems, yet their voice is limited. Unlike municipalities, metro areas made up of two or more contiguous municipalities, typically do not have clearly defined borders or unique legal standing. No politician speaks for Canada's metro areas. No urban practitioner is employed by a metro area.

An assessment of cities requires a metropolitan perspective, despite their limited voice. At the 2021 census, Canada had 41 census metropolitan areas as defined by Statistics Canada (CMA—see Section 1.10 for a discussion on CMAs) and 11 census agglomerations (CA). In 2021, 72% of Canada's population lived in a CMA. An additional 12% lived in a CA. The remaining 16% lived in rural communities. Statistics Canada regularly defines CMAs and highlights key trends within these urban areas, yet these changes are aggregates over several municipalities.

The largest city in a CMA is often only a fraction of that metropolitan area's total population. The mayors of Vancouver, Montreal, and Toronto represent less than half the city's total CMA population. The main city (municipality) in Canada's eight largest CMAs makes up the following share of the total CMA population: City of Toronto—45%; City of Montreal—41%; City of Vancouver—25%; City of Ottawa—68%; City of Calgary—88%; City of Edmonton—71%; Quebec City—65%; Winnipeg—90% [27].

Even though the eight largest CMAs alone are home to half the country's population and generate more that 60% of GDP, and that the CMA as a unit of analysis is the most important, no one speaks for Canada's CMAs. The limitation is even more pronounced in Toronto and Vancouver, where contiguous but differentiated CMAs, such as Oshawa, Hamilton, and Abbotsford, are integral parts of the broader metropolitan economy. As critical as the urban economy is to Canada, and that the efficient functioning of metropolitan services and productivity is Canada's most important driver of wealth, very few politicians (at any level), citizens, businesses, or urban practitioners would be able to identify the borders of their respective CMA. The most important economic units of Canada have no unique voice.

Giving voice and agency to Canada's key metro areas, including CMAs, is likely the most powerful way to improve national productivity and meet sustainability goals (see Chap. 6). Canada's metro areas and CMAs are unlikely to hold legal standing or exercise independent agency. They will however increasingly exhibit agentic behavior as circumstance and opportunity arise.

A few other trends are worth noting that attempt to give voice to systems and peoples with limited voice of their own. These include Councilors for the Future and the recent legal standing for Muteshekau Shipu in Quebec (the Magpie River) [28]. In Ontario, a youth-led climate case (Mathur et. al) is arguing that insufficient efforts have been made toward the fight against the 'climate emergency in Canada' [29]. Similarly, youth in Montana, USA, joined more than 30 legal challenges around the world [30]. These cases of intergenerational equity have been affirmed in the states of Washington and Massachusetts, and in Germany, Ecuador, Colombia, and Belgium. In 2017, a local Māori tribe in North Island, after 140-years of negotiations, received legal status for the Whanganui River—the third-largest in New Zealand [31].

There is influence through a collective voice; however, Canada's municipal associations are skewed toward rural representation. The Federation of Canadian Municipalities (FCM) has 2000 members, but more than 80% of the members are rural municipalities (even though they contain less than 20% of the country's population). This trend is repeated provincially. For example, of the 126 members of the Association of Municipalities of Ontario (AMO) only about 30 represent cities (and the City of Toronto is not a member).

Humans prevailed over other species because of their ability to cooperate, civilize through cities, and share narratives that facilitate scaling and efficiencies of networks and platforms. Laws and currencies, nationalities and cultural norms are all intangible things brought about through shared narratives [32]. These narratives are not necessarily complete or true—rarely is one person, or a city, or a country, correct on everything. Recognizing this fallibility, the metropolitan voice and sustainability mindset broaden the ability of cities to cooperate across regions and countries.

2.3 The Century of Countries

In 1945, when the United Nations was established, there were about 75 sovereign states; [33] the initiating charter of the UN was signed by representatives of 50 countries on June 26, 1945 (Poland signed later in October). Since then, membership swelled to 193.

Since 1920, 122 new countries emerged or re-emerged from colonial rule, or split off from larger countries (50 in Africa; 32 in Asia; 15 in Europe; 13 in the Americas; 12 in Oceana) [34]. Very few countries consolidated. Hawaii joined the USA and Newfoundland and Labrador merged into Canada. Much of the last 100 years has been preoccupied with countries defining themselves, establishing institutions and alliances, and developing new frameworks for cooperation, or antagonism.

The number of countries participating in the modern Olympics is illustrative. Twelve countries participated in the 1904 Olympics; 22 countries participated in 1912; and 29 participated in 1920. Compare this to the 2016, 2020, and 2024 Summer Olympic Games that each had a record 206 countries participating.

The populations of cities ebbs and flows. In Canada, overall, cities are still flowing, adding one to two million people per year, depending on immigration rates. Rates of growth (and decline) vary across regions and cities. For example, in 1920 Montreal was the largest city in Canada. Winnipeg was the third largest.

As the world further urbanizes, the era of city growth that started accelerating in earnest in 1950 continues. With 45 cities above one million in 1950 increasing to 400 in 2000, that number is expected to more than double again by 2050 (to 803, of which six are in Canada). By 2100, the number of cities above one million population worldwide is expected to be around 955 (of which seven will be Canadian). Urbanization is dramatically increasing the size of secondary cities, those under one million population. In 1950, the world's total urban population was 746 million. This increased fourfold to 2.95 billion in 2000 and is expected to more than double again to 6.39 billion in 2050 and exceed nine billion by 2100.

As Table 2.1 illustrates, global wealth increased in lockstep with the world's rise in urban populations. From 1900 to 2000, the average per-person wealth increased almost tenfold (by GDP). The Great Acceleration that gained speed in the 1950s driving a rapid increase in energy and material use (Chap. 5), brought with it a similar acceleration of wealth and waste. People got much richer through cities, but most ecosystems were highly impacted. Greenhouse gas emissions and a rapidly warming climate are one powerful example.

2.4 An Uneven Relationship

Trade is more important to Canada's economy than just about any other major country. Trade makes up 62% of Canada's GDP (30.9% exports; 31% imports) while trade makes up just 26% of GDP in the USA (10.9% exports; 14.6% imports) [35]. Eighty percent of Canada's trade is with the USA. It's one of the world's largest trading partnerships, but is very lopsided. Trade with the USA is about a third of Canada's economy whereas it is barely 3% of the US economy.

Michigan is the most trade-intensive state with Canada but trade with Canada represents only 11% of Michigan's economy. Compare this to the share of provincial economies trading with the USA: 39% of Ontario, 46% of Alberta, and 62% of New Brunswick. Trade with Canada is less than 2% of New York's economy, and less than 1% of either California or Florida [36].

Despite the Great Lakes Region hosting the largest trade flow between any two countries, and a robust trade corridor in Cascadia stretching from Vancouver to northern California, Canada is vulnerable to disruptions in trade with the USA. Irritants, such as Canadian agricultural protections, softwood lumber, illegal immigration, smuggling contraband, and transmission and pipelines, along with political vicissitudes, suggest the need for trade regionalization to augment North American trade agreements [37].

As populations age, their overall demand for manufactured goods declines. Canada's cities will encourage greater emphasis on the trade between services and people. Intellectual property, credentialization, and e-commerce take on new importance. Cities are as concerned with agreements that facilitate the trade of goods and services as they are with levels and capabilities of immigrants (and increasingly, emigrants). Canadian partnership agreements (trade and immigration) are likely to focus on the Indo-Pacific region where two-thirds of the world's middle class will soon be located. Africa, with its burgeoning cities, will also emerge as an important region.

The old, rigid model of professionals like doctors and nurses, and increasingly tradespeople, picking up roots, severing ties, and immigrating from countries like Nigeria and Philippines, is likely to shift into a more fluid partnership. The shift from oil and gas as the key commodity traded globally, to critical minerals, as the world decarbonizes, plus aging populations in many countries, provides an opportunity (as well as a threat) to Canada to reform trade and immigration. Cities like Vancouver, Calgary-Edmonton and Toronto may have different objectives from Canada as a whole.

2.5 The Power of Geography

Cities possess several unique attributes. They are fixed in place. In the last 100 years, almost every country had its border shift, yet cities have a unique *querencia*, the place where they are anchored. The terminus of a portage, a safe harbor, or where two or more roads or railways meet, cities emerged where geography dictated.

Toronto was first settled by the Iroquois in the village of *Teiaiagon* on the east side of what is today the Humber River, along the Toronto Carrying-Place portage [38]. Montreal, Saint John, Quebec City, along with London, Rotterdam, New York, Jakarta, and Shanghai, all owe their genesis to favorable geography.

This introduces the greatest vulnerability of cities; they are unable to get out of harm's way. Sea level rise, regional and local heat islands, flooding and droughts mean cities will need to adapt and harden their infrastructure. This will introduce even greater tensions within countries as citizens in general may think that strategic retreat for a specific city may be a better response for the overall country, while residents of that specific city may believe differently.

Small villages along the shores of Lakes Ontario, Erie, Huron, Michigan, and Superior and next to the St. Lawrence River grew into some of Canada's and the USA's largest cities. As the communities grew, and trade among them intensified, they formed the world's largest megalopolis. About 110 million people live in the Great Lakes Region. If the region were to be its own country it would have the world's third-largest economy, behind only the USA overall, and China. Cascadia, a similar megalopolis is growing along the Pacific Northwest and includes the cities of Vancouver, Seattle, and Portland.

2.6 The City as a Wicked Problem

A city is a living system and therefore has agency. This agency is not always easy to express and is often overshadowed by a province or nation. A city's agency is a culmination, or the mean of all residents, institutions, and businesses. And individual agency is rarely equally applied. In Canada, along with most high-income countries, inequality is rising. The voice of the poor is declining.

Bigger businesses retain lobbyists largely to protect the status quo. And those same businesses and institutions are adept at exerting influence on provincial and national policies, which often govern for an area larger than a single city. Residents in other regions have a strong voice in shaping the policies of provincial and national governments, often out of sequence, or at odds with city residents.

In many countries, provincial (or state) and national elections see voter preferences bifurcate strongly along regional and rural–urban lines. Canada's 2022 Parliament for example: Of the 157 seats that brought the Liberal party to power after the September 2021 election, 80% were held in Vancouver, Toronto, Ottawa, and Montreal. Regionally, Alberta with two of the country's most important cities, Calgary and Edmonton, had just one Liberal seat.

The ability to exert agency by cities is further diminished in many countries, especially in Canada, as rural voters tend to have a higher per ballot influence than their urban counterparts. The input from most city residents is muted—even in voting, only about a third of the community expresses their wishes in who would lead their city.

"*El pueblo unido jamás será vencido*" (the people united will never be defeated) is a phrase, and song, popularized by the progressive social movements in Chile in the early 1970s. The people in cities give rise to social movements. These social movements underpin the complexity of cities. Obtaining consensus across a community is difficult, and even more-so when the consensus may be shifted along broader provincial, national, or increasingly, international scales.

> That which is not good for the beehive cannot be good for the bees.
> —Marcus Aurelius

The strength of bees is their ability to cooperate and coordinate. In attempting to address the aspirations and abilities of people in a city, or system of cities, as well as the sustainable operation of urban services and community well-being, the challenge of coordination arises. This was borne out in Canada's response to COVID-19. In *Resilient Institutions: Learning from Canada's COVID-19 Pandemic*, the paramount need for coordination within and between agencies and institutions was raised [39]. This failure of coordination during COVID-19 was emphasized in the series of articles and editorial of the prestigious British Medical Journal *The world expected more of Canada* [40].

When reviewing urban services in Toronto Region, the Task Force on Greater Toronto Area services (The Golden Report) came to a similar and inescapable conclusion: underlying nearly all the problems and issues facing the city-region were a fundamental lack of coordination [41].

Managing a pandemic or a metropolitan region is akin to addressing a wicked problem, which is inherently difficult to solve for three reasons: (1) it involves interconnected economic, cultural, and social factors; (2) it tends to be long-term in nature; and (3) its possible solutions can be contentious due to entrenched thinking and interests.

Managing a city, especially a large city, subject to varying pressures from agencies, and community groups, is a wicked problem.

Characteristics of a wicked problem include [42]:

- There is no definitive formulation of a wicked problem.
- Wicked problems have no stopping rule; i.e., there is no point in time at which the process of addressing a problem is completed.
- Solutions to wicked problems are not true-or-false, but good-or-bad.
- There is no immediate and no ultimate test of a solution to a wicked problem.
- Every solution to a wicked problem is a 'one-shot' operation.
- Wicked problems do not have an enumerable or exhaustively describable set of potential solutions, nor is there a well-described set of permissible operations for addressing wicked problems.
- Every wicked problem is essentially unique.
- Every wicked problem can be a symptom of another problem.
- The analyst's world view is the strongest determining factor for explaining differences in descriptions of wicked problems and preferences for how they should be addressed.
- The planner has no right to be wrong.

2.7 Intervening in Urban Systems

System analyst Donella Meadows (and lead author of *Limits to Growth* [43]) suggested we cannot impose our will on a system. However, we can listen to what the system tells us and discover how its properties and our values can work together to bring forth something much better than could ever be produced by our will alone.

Meadows proposed leverage points to intervene in a system. For cities, these can loosely be categorized as either governance or management interventions as urban systems need to be governed *and* managed. Occasionally the two categories overlap, but not often. The impact of interventions varies with governance interventions such as developing a new narrative or paradigm having a much greater impact than management indicators like setting targets and providing credible metrics. However, the ease of implementation typically varies inversely with impact. Management interventions should be easier to implement than those based on values and a shared narrative.

Broadly speaking *governing* the system is discussed and debated by many, while *managing* is left to others. Arguably in Canada, like most countries, the national government provides more governance than management, while local governments, most of whose leaders in Canada remain non-partisan, tend to emphasize management. This view might be corroborated through voter turnout in Canadian elections. Typically, about 65% of eligible voters cast ballots in national elections [44], while voter turnout for municipal elections is much lower. For example, turnout was about 36% in recent Ontario and BC municipal elections. Provincial elections tend to vary between 50 and 65% voter turnout, with higher participation in Quebec. Ontario had a record low voter turnout of 44% in 2022.

Countries typically are motivated through a common, passionate narrative that influences governance. For example, a secular, cultured France, freedom in the USA, Russia as a distinct civilization, and a 'Confucian' China stabilizing the world order.

Canada, in addition to the narrative of abundance, emphasized peace, order, and good government. Federal Governments however stoked (and responded to) national fervor through the CP Railway, emphasis as a 'middle power' supporting multilateralism, and national health care.

Cities rarely had their own narrative. Montreal provides a powerful example where the city was caught in the passions of provincial separatism as corporate head offices moved away and English-speaking universities see enrollments drop. Jane Jacobs argued that Montreal would have benefited from Quebec seceding from Canada, highlighting the strength of paradigms and narratives in shaping cities [45, 46].

Diverging narratives between provincial values and city aspirations are apparent in Calgary-Edmonton and Alberta, and increasingly Toronto Region and Ontario. Cities will continue to focus foremost on management aspects of system interventions; however, larger urban areas are likely to place greater emphasis on the narrative and setting goals for system interventions.

Donella Meadows suggested twelve places to intervene in a system [47]. In descending order of impact and increasing ease of implementation, the interventions can be adapted to cities (urban systems) as listed below.[1]

Governance (Community Narrative)

i. Ability to transcend paradigms. Akin to enlightenment and self-actualization—going beyond challenging fundamental assumptions, into the realm of changing the constructs that fostered the original assumptions. For example, the ability to rise above, and integrate, two (or more) disparate worldviews such as First Nations mother earth and capitalist market dominance.

ii. City paradigm. A shared idea (values) in the minds of citizens, e.g., Rousseau's social contract. Thomas Kuhn's Structure of Scientific (System) Revolutions suggests working with active change agents and more open-minded, middle-ground citizens, e.g., abolishing slavery, colonialism [49].

---

[1] Adapted from [48].

Governance and Management

iii. Goals of the city. Broad, system-level goals such as survival, resilience, differentiation, and evolution. City goals usually align with upper-tier governments; however, if not, rationalization of goals and objectives is needed, and selection of paramount goal(s).
iv. Ability to self-organize. Society's capacity to innovate and adapt to changing circumstances (and objectives), applied human creativity, ability to surpass system constraints.

Management (Sustained High-Quality Service Delivery)

v. Rules of the system (city). Who makes the rules (e.g., laws, regulations, standards), who enforces them (and how), and what is the mechanism to change the rules.
vi. Information flows. Better provision of information (timeliness, completeness). Increased accountability.
vii. Gains from positive feedback loops. This can lead to unconstrained growth and increased inequality through 'over-heated' economy (self-reinforcing), e.g., nutrient loading in a lake and eutrophication.
viii. Strength of negative feedback loops. The use of preventative medicine and maintenance, full-cost accounting, a price on pollution, 'fast-track' construction, and 'just-in-time' delivery (self-correcting).
ix. Lengths of delay, responsiveness. Time to build can have a significant impact, however often difficult to 'fast-track'.
x. Built structure and nodes of intersection. 'Lock-in' effects critical, as costs to change significantly higher than 'building it right' initially (with keen emphasis on optimum flow—e.g., people, traffic, energy, water).
xi. Buffering capacity and urban resilience. The system's (city's) ability to stabilize and ameliorate potential shocks, perturbations, and supply disruption.
xii. Constraints, parameters, targets, and operating parameters. Metrics are typically well known by citizens but provide little ability to bring about behavioral change.

2.8 The City as Commons

In 1968 George Hardin wrote his *Tragedy of the Commons* [7] arguing that the planet, 'like a lifeboat', had provisional limits, and that there are areas where, if left unchecked, people would overuse resources, leading to collapse. Farmers adding yet more sheep to an already over-taxed common pasture, or fishers, taking more fish than can possibly be sustained. The *city as commons* suggested similar potential tragedies from overuse of common areas. Traffic congestion providing a powerful example.

2.8 The City as Commons

The city as commons also, however, suggests opportunity. Everyone can be entitled to the city and build on the opportunities provided. The city as commons offers opportunity and promise for all.

Elinor Ostrom was awarded the Nobel Prize in Economics for her work on managing the commons [6]. This work initially focused on more rural commons such as shared fishing stocks, forests, and grazing lands. However, toward the end of Ostrom's career she focused on the city as commons. Ironically on the day of her death, she published an Op-ed extoling the virtues of cities and their ability to bring from the grassroots, a common narrative and shared platform for action (see Sect. 2.15). Ostrom is credited with providing a set of principles for managing the commons. These can readily be applied to managing the commons of a city (see below).

In Canada, a practical rule of thumb for a city's hinterland is the airport typically used for international travel. In Toronto and Vancouver, the urban agglomerations of the Great Lakes Region and Cascadia emerge, with airports in Buffalo (and Detroit) and Seattle less than a three-hour drive away. The role of Toronto *vis-a-vis* all of Canada is also highlighted through its Pearson Airport that handles almost half of the country's overall international air traffic (see Sect. 3.3).

Satellite cities such as Victoria and Kamloops to Greater Vancouver; Sherbrooke, Trois-Rivieres, Cornwall, and Ottawa to Montreal; Belleville, Peterborough, Barrie, and London to Toronto, and farther out, Windsor and eventually Sudbury with the completion of Highway 400, and North Bay with reinstatement of Northlander train service, are key communities of the broader urban area.

Edmonton and Calgary, along with Red Deer and other communities along the 300 km of Highway 2 connecting the two cities, are also emerging as a larger metro urban region in Alberta. In this book, Calgary-Edmonton is discussed as one large urban area. The two cities might be divided, in which case, six larger urban areas would be proposed rather than five.

Cities like Regina, Winnipeg, and Saint John are near the US border; however, they are not likely to emerge as parts of broader urban regions as they may be below the threshold of size and proximity to a larger US city.

Much of Canada's urban (and overall) future rests on the regions of Vancouver, Calgary-Edmonton, Toronto Region, Ottawa-Gatineau, and Montreal. These five urban commons support about 70% of Canada's GDP, a share that is increasing.

The common good of an urban area differs from the commons more typically ascribed to oceans and forests or common grazing areas for livestock. The services, or goods, that urban areas provide are less about natural resources and more about opportunity and well-being. The right to education, peaceful association, health care, housing, and a means to meaningfully participate in society and the economy, these are the common goods of a city.

History does not repeat itself, but it often rhymes [50]. Patterns emerge and repeat, and scale. So too in nature. This replication is referred to as *fractals*. The way water rushes down a drain scales up to the path of clouds and water in a hurricane; the branches of a stream follow the pattern of the river, and a small twig is repeated in the whole tree; the circulatory system of humans has an uncanny resemblance to tributaries and rivulets forming into a large river network.

Cities, the most organic and life-like structure of human civilization, replicate nature's patterns. Cities grow through networks, which are patterns of relationships. The way traffic bunches at a constriction follows the same pattern of water passing through a narrowing channel. Or the way people walk the shortest path they can. Humans walking through a city travel the same way a deer might walk through the woods or caribou across the plains. Plotting people moving in a large city resembles airways of the lung, the branches and roots of a mighty tree, or a large river's tributaries and channels—these patterns are often indistinguishable. These 'nature repeats', and nature-based, or nature replicated solutions, are an additional tool to help manage the city as a commons.

Principles for Managing a Commons

- Where practicable use natured-based approaches and replicate nature, e.g., biomimicry.
- Define clear group boundaries.
- Match rules governing use of common goods to local needs and conditions.
- Ensure that those affected by the rules can participate in modifying the rules.
- Make sure the rule-making rights of community members are respected by outside authorities.
- Develop a system, carried out by community members, for monitoring members' behavior.
- Use graduated sanctions for rule violators.
- Provide accessible, low-cost means for dispute resolution.
- Build responsibility for governing the common resource in nested tiers from the lowest level up to the entire interconnected system.

2.9 Canada, Growing Trouble

Canada's 200-hundred-year urban history from 1920 to 2120 will likely be marked by two halves of varied growth. During the first 100 years, from 1920 to 2020 Canada's GDP grew by an average of 2.7%.[2] This idea of relatively stable growth permeated into increasing housing prices, labor negotiations, and overall price increases. The second half, 2020 to 2120, will likely see noticeably lower growth, around 1%. The OECD, for example, projects that Canada's productivity growth to 2060 will be the lowest (around 0.8%) among advanced economies [52].

Collectively, Canadians are conditioned to higher growth than they are likely to see in future. With growing discontent as growth slows, public policies, such as a price on carbon, or welcoming immigrants, who may be seen as contributing to

---

[2] Based on 2.6 percent 1961 to 2022 as provided Government of Canada 2023 fall economic statement, and 3 percent estimate from GDP per capita, 1910 to 2018 [51].

housing price increases or rising unemployment, may be more challenging. Growth is also likely to vary regionally, exacerbating regional tensions.

Countries require narratives to keep people together: A shared language, history, and culture. 'A better future' often served as a rallying cry for social cohesion. A country provides currency and nationality and the means to participate globally, most of our laws, and hopefully our security and recompense if wronged. With the promise of a better future (financially) more uncertain, centripetal forces may play a larger role, leading to fracturing regions.

Cities also stir imaginations with rivalries like the Toronto Maple Leafs and Montreal Canadiens, or the Calgary Flames and Edmonton Oilers. The annual "Banjo Bowl" between Regina (Saskatchewan Roughriders) and Winnipeg (Blue Bombers) is a lively prairie competition. Perhaps the ads in the Toronto subway extolling the virtues of living in Calgary engender a gentle clash between cities. And Toronto often receives a frosty reception outside the GTA. However, these rivalries are much more benign than typical international tensions.

Several future peaks for Canadian cities are reasonably clear. By 2120, all cities in Canada are likely to be declining in population, having reached peaks between the early 2000s (e.g., Maritimes and more isolated communities like Sudbury, Trois-Rivieres, and Estevan) and the late 2000s (Toronto, Vancouver, Calgary, and Edmonton). Higher rates of immigration may delay these peaks, however with global population expected to peak around 2080, Canada's population will likely join most of the other OECD-member countries and be declining in the latter quarter of the twenty-first century.

With stable, or declining populations, energy and material consumption will also have reached their peaks and be in decline by 2120. Canada's relative global economic decline will have continued largely unabated from its zenith in the 1970s.

Looking back 100 years, a few large trends are clear. Toronto surpassed Montreal as Canada's largest city in 1971. Montreal's relative decline, *vis-a-vis* Canada and the rest of the world, was aggravated by sovereignty aspirations and assertion of French language rights.[3] Winnipeg, more remote cities, and those in the Maritimes, declined in population relative to burgeoning Calgary, Edmonton, and Vancouver.

At the turn of the twentieth century about a third of households were non-census members. A boarder, lodger or employee was common. Today less than 10% of householders are non-related, non-census family members. The common living arrangements with non-family members was brought about largely by deaths of family members. In 1921, about 9% of children aged 15 and under had experienced the death of at least one parent. Going forward, Canadian households may again have multigenerational, split-family members living together.

Perhaps there will always be friction between some countries as they struggle over resources or conflict over values. Nations will never be fully united as they feel the need to fight over access to energy, minerals, food, water, and increasingly, people. There may always be a country that Canada considers a threat.

---

[3] For example, the more recent reduction of provincial financial support to Concordia and McGill universities. Also, see [53].

False narratives abound. For example, the urban–rural conflict. Despite signs sprinkled about the countryside claiming "farmers feed cities" or anger directed at "cidiots" that retreated to their cottages during COVID, overwhelming local services, there is mutual attraction and dependence between the two groups. City residents pay the farmers for their work. In some cottage-country areas more than half the local taxes are paid for by city-folk (who often use less of those services). Both 'sides' benefit. Perhaps the best way for Canada to address declining economic growth is to enhance the growth of civility. Like fractals, this behavior will grow outward.

2.10 The Measure of Cities (Census Metropolitan Areas and Agglomerations)

Statistics Canada defines a census metropolitan area (CMA) or a census agglomeration (CA) as being formed by one or more adjacent municipalities centered on a population center (known as the core). As defined by Statistics Canada, a CMA must have a total population of at least 100,000 based on data from the current Census of Population Program, of which 50,000 or more must live in the core based on adjusted data from the previous Census of Population Program. A CA must have a core population of at least 10,000 also based on data from the previous Census of Population Program.

To be included in the CMA or CA, other adjacent municipalities must have a high degree of integration with the core, as measured by commuting flows derived from data on place of work from the previous Census Program.

Census metropolitan areas (CMAs) and census agglomerations (CAs) are large, densely populated centers made up of adjacent municipalities that are economically and socially integrated.

According to the 2021 Census, 84% of Canada's population lives within a CMA or CA. This amounts to over 31 million people. More than half of the population, some 20.5 million people, lives in the ten largest CMAs.

Once a population center attains a total population of 10,000 people, it is eligible to become the core of a census agglomeration (CA). Once a population center attains a total population of 50,000 people, then it is eligible to become the core of a census metropolitan area (CMA). The boundaries and population data for the cores that are used to delineate CMAs and CAs are taken from the previous census.

Since CMAs and CAs are based on census subdivisions (CSDs), a 'delineation core' is created from those CSDs that have at least 50% of its population living in the core.

Using commuting data based on the place of work question from the previous Census Program, commuting flows are calculated for workers going to the delineation core. If a surrounding census division (CSD) has a minimum of 100 commuters going into the delineation core and at least 50% of the employed labor force living in the

CSD works at a fixed workplace address in the delineation core, then the CSD is included in the census metropolitan area (CMA) or census agglomeration (CA).

This book proposes five larger urban regions in Canada, comprising one or more CMAs. These include Montreal (CMA), Ottawa-Gatineau (one combined CMA spanning two provinces), Toronto Region (comprising several CMAs, aka Golden Horseshoe), Calgary-Edmonton (combining CMAs and CAs along the Highway 2 corridor), and Vancouver (two CMAs).

2.11 The Impact of Humans: A Geological Perspective

In 1835, prior to visiting the Galapagos Islands, Charles Darwin experienced a major earthquake in Chile; a once-in-a-100-year chance. Prior to signing on to the Beagle, Darwin was already studying geology at Cambridge University. However, witnessing first-hand a 6.5-m shift of the earth's crust gave Darwin an understanding of the role of time in earth systems. Seeing direct mountain building helped him challenge the religious dogma that believed the earth was only some 6000 years old. This appreciation of the scale of time also underpinned his thinking on natural selection [54].

In 'thinking like a mountain' time is the key ingredient. Canada's youngest mountains for example, are the St. Elias Mountains, a subgroup of the Pacific Coast Ranges. Mount Logan, in Yukon Territory, Canada's highest peak, began forming some 35 million years ago as the smaller Yakutat tectonic plate, moving at a brisk 50 mm per year, slipped underneath the much larger North American Plate. Mount Logan at a current elevation of 5959 m is still undergoing active tectonic uplift of about 0.35 mm per year, slower than the rate a fingernail grows.

The Pacific Coast ranges are half as young as the Rocky Mountains that were formed between 80 and 55 million years ago. However, the Rockies are geologically youthful compared to Canada's oldest mountains, the Laurentians. The Laurentians of Quebec and Labrador are over 1 billion years old, making up a key part of the Canadian Shield, one of the oldest formations on Earth.

The human mind has a hard time appreciating the age of the cosmos, and how if the 4.5-billion-year span of planet earth were compressed into a 24-h day, dinosaurs appear at 10:56 pm, the first mammals at 11:39 pm, and the earliest humans just a little before 11:59 pm. The world's first cities appeared only 15 s ago, and Canada's oldest city, Saint John, NB, incorporated May 18, 1785, would appear less than two seconds before midnight.

Canada is an indomitable and vast land, with some bedrock older than a billion years. A 200-year horizon as suggested in this book is a geological blip, yet much longer than most political considerations or the lifespans of businesses. The 1920–2120 period discussed in this book is likely to be the most germane regarding the impact of humans on planet earth. By 2120 most planetary systems should start to recover from the impacts of humans. Tipping points may have been triggered and several millennia will still be needed to return the atmosphere and oceans to relative stability. However, with human populations declining and greenhouse gas emissions largely abated, recovery from the Great Acceleration will be underway.

Canadian geomorphology was mostly shaped by the enormous impacts of glaciers. As the 1-to-2-km-thick ice sheets scoured the landscape they moved entire mountains of earth and stone. Rivers, with their inexorable erosion, gave rise to huge land masses like the McKenzie and Peace-Athabasca Deltas. However, beginning in 1945 humans surpassed the natural processes of wind and water in moving earth and aggregate. Today humans produce and move at least 350 Gt (165 km3) of sediment, an amount more than 27 times greater than the sediment moved by all the world's rivers [55].

Krausmann et al. [56] corroborate this estimate of the massive movement of material stocks for buildings, infrastructure, and machinery, in their paper outlining how global material stocks have risen significantly over the twentieth century. About half of the world's annual resource extraction is used to build and service our cities, some 1200 billion tons a year.

With more than four billion years of planetary history, earth scientists rarely can claim something is unprecedented. Even if limited to the last 500 million years when complex life flourished across the planet, there have been five mass extinctions (where more than 75% all species are lost over a relatively short time frame—a 'relatively short' time frame is defined as less than 2 million years). The planet is now entering the sixth mass extinction.

Scientists do not believe we are about to replicate the Permian–Triassic extinction ("The Great Dying" where almost 90% of species became extinct [57]) however the *rate* of planetary change is truly unprecedented. The human species is likely not to go extinct from the current planetary perturbations we are causing. However, the scale of planetary change and overall degradation is alarming.

At the start of the Holocene, some 11,700 years ago, atmospheric CO_2 concentrations were about 270 ppm (i.e., 0.027%). Methane (CH_4) concentrations varied but were around 600 ppb (parts per billion, i.e., 0.00006%). Today concentrations of CO_2 are around 425 ppm (a 156% increase and rising about 2.5 ppm per year); methane, a more potent greenhouse gas, is now over 1920 ppb (a 320% increase). Annual increases of methane concentration vary, but are around 17 ppb, and rising [58].

Geologically speaking the increase in greenhouse gases is enormous and relatively instantaneous to planetary systems. A mass extinction that provides a precedent for what earth is now undergoing is the Cretaceous-Paleogene extinction, about 66 million years ago (which killed the dinosaurs). This is the only extinction event attributed to an asteroid hitting the earth. This is a comparable scale to the pace of planetary system change now underway with increasing atmospheric carbon dioxide (and other GHGs) and associated warming and ocean acidification.

2.12 The Shift to Sustainability

> If some countries have too much history, Canada has too much geography.
> —Prime Minister William Lyon Mackenzie King

Creating a great global civilization, where sustainability is understood and steeped through every aspect of society, is doable. Some would argue we are halfway there [59]. The next 100 years will see the Great Acceleration that began in earnest after WWII, slow and eventually decline as the global population peaks around 2085. For the first time all of humanity could see basic human needs, and most of our aspirations, met.

Canada's collective behavior emphasizes the country's history of resource development and much of the country's unsustainable nature. This resource-dependent culture permeates federal relationships with the provinces. Electricity generation and export is the largest component of Quebec's economy, Newfoundland and Labrador will add electricity export to oil and gas development (eventually making up about half the province's total economy).

Ontario's economy was initially turbo-charged with low-price electricity and automobile manufacturing (and designing communities to be highly supportive of auto-dependent mobility). Toronto particularly benefited from Canada's resource extraction economy with 40% of the world's mining companies listed on the Toronto Stock Exchange. Ontario, with more than 6000 active and abandoned mine sites, and a history of providing beaver pelts for Europe's fashions, is steeped in resource extraction.

Traveling west across the country, resources maintain their hold on provincial thinking. Agriculture, electricity export, oil and gas development, mining, and forestry, underpin the economies of Manitoba, Saskatchewan, Alberta, and British Columbia.

This resource reliance mindset filtered down to Canada's cities whose residents now live in the world's most resource intensive and wasteful communities. Per person, Canadians use more building materials, energy, and water, and generate more greenhouse gas emissions and solid waste than anywhere else on earth.

Globally a transition to sustainability is underway. This is most evident in the energy transition, where the world is collectively endeavoring to bring carbon emissions down to zero by mid-century. Additional efforts include shifting to a "circular economy", with more than the current rate of 7% of material reuse in the economy (global targets set at 50% by 2050). Global efforts to protect biodiversity and materials treaties, e.g., plastics, are also progressing.

Greenhouse gas emissions, stressed water bodies (e.g., from sediment, airborne pollutants, nitrogen, over-fishing), air pollution, and solid waste, are all by-products of resource extraction and use. Despite the world's growing affluence and urban populations, collectively the main goal of humanity over the next 100 years will be to use fewer resources, leading to much less damage of planetary ecosystems. Ideally bringing combined impacts below what can be sustained by the planet by the end of the century.

Perhaps ironically, the bulk of Canada's wealth generation today occurs not through resource extraction, electricity generation, or agriculture, but through the activities that take place in the country's urban areas. Fully two-thirds of the country's economy is generated within the cities (and two-thirds of that in the five largest urban areas). This urban share of the economy is growing.

A city's relationship with energy and materials is more nuanced than a business, province, or national government. Cities need to balance support to the resource businesses headquartered in the city. About 30% of employment in Calgary for example, is linked to oil and gas development, and about a third of the Toronto Stock Exchange is energy and resource-based companies. Toronto's longest and largest, continuous conference is held by the Prospectors and Developers Association of Canada (since 1932). Similarly, the Calgary Stampede is steeped in the city's oil and gas sector. Changing these events to a conservation mindset will be challenging.

2.13 The Power of Hierarchies—A Fable

Once upon a time there were two watchmakers. Let us call them Dana and Doug. They were both fine watchmakers and had many customers. People regularly dropped into their shops, and orders arrived by phone and email. Over the years, however, Dana prospered, while Doug's business languished. That is because Dana discovered the power of hierarchy.

Each watch made by both Dana and Doug included about one thousand parts. Doug assembled his watches in a way that if he had one partially assembled and had to put it aside for a break or an interruption it fell to pieces, and he had to start over at the beginning. The more Doug's customers contacted him, the harder it was for Doug to find uninterrupted time to finish a watch.

Dana's watches were no less complex than Doug's, but she put together stable subassemblies of about 10 parts each. Then she put 10 subassemblies into a larger assembly, and ten of those together made up a completed watch. Whenever Dana was interrupted and had to put down a partly completed watch, she only lost a small part of her work. Dana made her watches much more efficiently than did Doug.

Complex systems, like cities, evolve from simple systems only if there are stable intermediate forms. The resulting self-organizing complex forms will naturally be hierarchic. With time to evolve, forests, beehives, streams, and rivers, as well as communities will all develop and depend on hierarchies.[4] Cities, as humanity's most complex undertaking, are made up of hierarchies, and cities as systems of systems also evolve into hierarchies themselves, regionally, nationally, and internationally.

---

[4] Paraphrased from [60]; Simon quantifies the increased efficiency Dana receives through the application of hierarchies. If the probability (p) of being interrupted while assembling any one part is about 0.01 (1%) it will take Doug about 4,000 times longer to assemble a watch as Dana.

2.14 Networked Hierarchies and the Superpower of Cities

Ill-behaved little boys, and its usually boys, learn that if you take a stick to a termite mound, or large ant hill, and stir up trouble, the insects break down into smaller and smaller groups. They work together in small groups to repair the damage. As repairs are carried out, the groups slowly re-connect. Sometimes it can take hours or days, but the community recovers and re-forms into the larger single group.

This same behavior is observed when cities take a blow. A hurricane, flood, fire, social unrest, or earthquake disrupts the networks, and people are forced to fend for themselves in smaller groups. These households and neighborhood clusters coalesce into larger groups and eventually re-form into pre-existing networks and institutions. In a city with higher population densities, these networks, proximity, and agglomeration effects tend to override the natural human behavior of forming into tribes and cultural groups. Almost all cities have ethnic enclaves. Chinatown, Little Italy, Greektown, neighborhoods with higher shares of an ethnic group are common. However, in all Canadian cities these neighborhoods have relatively permeable borders, and the networks and institutions of urban service delivery transcend these single areas, requiring linkages.

2.15 Green from the Grassroots: Elinor Ostrom Championing the Power of Cities

In the lead up to the Rio + 20 UN Conference Elinor Ostrom's last commentary was published on June 12, 2012, the same day of her death. Ostrom argued that those trying to seek a global agreement on sustainability should recognize that diverse actions at the city and regional level are most needed for systems challenges like climate change.

"We have never had to deal with problems of the scale facing today's globally interconnected society. No one knows for sure what will work, so it is important to build a system that can evolve and adapt rapidly.

Decades of research demonstrate that a variety of overlapping policies at the city, subnational, national, and international levels are more likely to succeed than are single overarching binding agreements. Such an evolutionary approach to policy provides essential safety nets should one or more policies fail.

The good news is that evolutionary policymaking is already happening organically. In the absence of effective national and international legislation to curb greenhouse gases, a growing number of city leaders are acting to protect their citizens and economies.

This is hardly surprising—indeed, it should be encouraged."

Ostrom's plea for urgency of action has so far not been heeded.

"The goal now must be to build sustainability into the DNA of our globally interconnected society. Time is the natural resource in shortest supply, which is why the Rio summit must galvanize the world. What we need are universal sustainable-development goals on issues such as energy, food security, sanitation, urban planning, and poverty eradication, while reducing inequality within the planet's limits" [61].

The Rio + 20 conference facilitated the transition from the Millennium Development Goals (MDGs) to the Sustainable Development Goals (SDGs) that were unanimously adopted September 25, 2015, by the 193 countries of the UN General Assembly. The SDGs continue to be buffeted by events, and most will not be achieved by the 2030 target date. However, the SDGs have emerged as an excellent means to measure success in how well a community, or the planet collectively, is building sustainability into the DNA of our globally interconnected society. As suggested by Elinor Ostrom, the next iteration of development goals, post-2030, will likely be scalable and designed to start from cities and grow from the grassroots.

References

1. Clark G (2016) Global cities: a short history. Brookings Institution Press
2. Bettencourt LM (2013) The origins of scaling in cities. Science 340(6139):1438–1441
3. West G (2017) Scale: the universal laws of life, growth, and death in organisms, cities, and companies. Penguin
4. Batty M (2013) The new science of cities. The MIT Press
5. Meadows DH (2008) Thinking in systems: a primer. Chelsea Green Publishing
6. Ostrom E (1990) Governing the commons: the evolution of institutions for collective action. Cambridge University Press
7. Hardin G (1968) The tragedy of the commons. Science 162(3859):1243–1248
8. NewWorldEcyclopedia.org (2024)
9. Encyclopedia Britannica (2024) Beijing
10. Hoornweg D (2018) Sustainability through the world's cities. In: 1st international conference on new horizons in green civil engineering (NHICE-01), Victoria, BC, Canada
11. Xu C, Kohler TA, Lenton TM, Svenning JC, Scheffer M (2020) Future of the human climate niche. Proc Natl Acad Sci 117(21):11350–11355
12. World Bank (2009) World development report
13. Alvarez J, Krznar MI, Tombe T (2019) Internal trade in Canada: case for liberalization. International Monetary Fund
14. Deloitte (2021) The case for liberalizing interprovincial trade in Canada.
15. Dumais M (2012) The negative consequences of agricultural marketing boards. Montreal Economic Institute. See also: Cardwell R, Lawley C, Xiang D (2015) Milked and feathered: the regressive welfare effects of Canada's supply management regime. Canadian Public Policy 41(1):1–14
16. Elliot T, Goldstein B, Charlebois S (2025) Over 6 billion liters of Canadian milk wasted since 2012. Ecol Econ 227:108413
17. Jacobs J (1961) The death and life of great American cities. Vintage Books
18. Weaver W (1948) Science and Complexity. Am Sci 36(4):536–544
19. Bettencourt LMA (2013) The kind of problem a city is: new perspective on the nature of cities from complex systems theory (SFI working paper: 2013–03–008). Santa Fe Institute
20. Dong L, Duarte F, Duranton G, Santi P, Barthelemy M, Batty M, Ratti C et al (2024) Defining a city—delineating urban areas using cell-phone data. Nat Cities 1(2):117–125
21. Bejan A, Lorente S (2011) The constructal law and the design of the biosphere: nature and globalization. J Heat Transf 133
22. Bejan A, Peder Zane J (2012) Design in nature. Doubleday

23. Bettencourt L, Lobo J, Helbing D, Kuhnert C, West G (2007) Growth, innovation, scaling, and the pace of life in cities. Proc Natl Acad Sci 104:7301–7306
24. Lobo J, Bettencourt L, Ortman SG (2024) Urban scaling theory: answers to frequent questions. Environ Plan B: Urb Anal City Sci 23998083241308418
25. Bettencourt L (2013) The origins of scaling in cities. Science 340:1438–1441
26. Tainter J (1988) The collapse of complex societies. Cambridge University Press
27. Statistics Canada 2021 Census
28. Canadian Geographic (2022) I am Mutehekau Shipu: a river's journey to personhood in eastern Quebec
29. Lumby C (2023) Acting against inaction: youths advocate for charter rights in the case of Mathur et al. (May 9, 2023). Canadian Law Review Research Paper No. 3, SSRN: 4443449
30. Donger E (2022) Children and youth in strategic climate litigation: advancing rights through legal argument and legal mobilization. Trans Environ Law 11(2):263–289
31. https://www.theguardian.com/world/2017/mar/16/new-zealand-river-granted-same-legal-rights-as-human-being#:~:text=In%20a%20world%2Dfirst%20a,an%20ancestor%20for%20140%20years (2024-08-28)
32. Harari YN (2015) Sapiens. Harper
33. Butcher CR, Griffiths RD (2020) States and their international relations since 1816: introducing version 2 of the International System(s) Dataset (ISD). Int Interact 46(2):291–308
34. Wikipedia. https://en.wikipedia.org/wiki/List_of_sovereign_states_by_date_of_formation Accessed 7 July 2024
35. World Integrated Trade Solutions data, World Bank. Accessed 7 July 2024
36. Tombe T (2024) The hub: is Canada betting too much on the U.S.?
37. The global exchange: the future of trade is regional. Canadian Global Affairs Institute, Podcast. Recorded 11 Apr 2024
38. Robinson PJ (2019) Toronto during the French Regime. University of Toronto Press, Toronto
39. Breton C, Han J, McLaughlin D, Woodward C (2024) Resilient institutions: lessons from Canada's pandemic response. Institute for Research on Public Policy
40. British Medical Journal (2023) 382:1634
41. Greater Toronto Area Task Force, Golden A (1996) Greater Toronto: background reports to the GTA Task Force. Queen's Printer
42. Rittel HWJ, Webber MM (1973) Dilemmas in a general theory of planning. Policy Sci 4:155–169
43. Meadows DH, Meadows DL, Randers J, Behrens III WW (1972) The limits to growth-club of Rome
44. https://www.elections.ca/content.aspx?section=ele&dir=turn&document=index&lang=e
45. Galbraith J, Goodman P, Jacobs J, Kierans E, King Jr M (2007) The lost massey lectures: recovered classics from five great thinkers. House of Anansi
46. Jacobs J (2016) The question of separatism: Quebec and the struggle over sovereignty. Vintage
47. Meadows D (2015) Leverage points-places to intervene in a system
48. Hoornweg D (2018) Sustainability through the World's cities in Mukhopadhyaya, P. (2018). NHICE 01 conference proceedings
49. Kuhn TS (1962) The structure of scientific revolutions. The University of Chicago Press
50. Usually attributed to Mark Twain
51. Our World in Data. Accessed 15 July 2024
52. Guillemette Y, Turner D (2021) The long game: fiscal outlooks to 2060 underline need for structural reform. OECD Economic Policy Papers, No. 29
53. Jacobs J (2007) The Lost Massey Lectures
54. Woodward K (2022) Aging in the anthropocene. Critical humanities and ageing: forging interdisciplinary dialogues
55. Cooper AH, Brown TJ, Price SJ, Ford JR, Waters CN (2018) Humans are the most significant global geomorphological driving force of the 21st century. Anthropocene Rev 5(3):222–229
56. Krausmann F, Wiedenhofer D, Lauk C, Haas W, Tanikawa H, Fishman T, Haberl H et al (2017) Global socioeconomic material stocks rise 23-fold over the 20th century and require half of annual resource use. Proc Natl Acad Sci 114(8):1880–1885

57. Encyclopedia Britannica (2024)
58. https://climate.copernicus.eu/climate-indicators/greenhouse-gas-concentrations Accessed 1 Aug 2024
59. Ritchie H (2024) Not the end of the world: how we can be the first generation to build a sustainable planet. Random House
60. Meadows D (2008) Thinking in Systems; Simon HA (1962) The architecture of complexity
61. https://www.project-syndicate.org/commentary/green-from-the-grassroots-2012-6#LfIxrb SgDt3u5ppe.99. Accessed 28 June 2024

Open Access This chapter is licensed under the terms of the Creative Commons Attribution-NonCommercial-NoDerivatives 4.0 International License (http://creativecommons.org/licenses/by-nc-nd/4.0/), which permits any noncommercial use, sharing, distribution and reproduction in any medium or format, as long as you give appropriate credit to the original author(s) and the source, provide a link to the Creative Commons license and indicate if you modified the licensed material. You do not have permission under this license to share adapted material derived from this chapter or parts of it.

The images or other third party material in this chapter are included in the chapter's Creative Commons license, unless indicated otherwise in a credit line to the material. If material is not included in the chapter's Creative Commons license and your intended use is not permitted by statutory regulation or exceeds the permitted use, you will need to obtain permission directly from the copyright holder.

Chapter 3
Jazz in the Kitchen: The Special Case of Toronto

If ever there was a case of too many cooks in the kitchen, it is Toronto. Simply defining Toronto is challenging, leave alone governing and managing it.

The Greater Toronto Area (GTA) is the minimum urban agglomeration that should be considered when planning service delivery, especially services like transportation, waste disposal, water supply, economic development, and social assistance. Often Hamilton, with a population of 625,000, is added to this urban region, making it the GTHA. Despite the increasing need, the 30 mayors and chairs of the GTHA, and their respective municipal staff, businesses, and residents, rarely agree, and when they do, it may well be the province or federal government deciding on the issue—often at odds with the local governments.

The challenge of orchestrating consensus in Toronto is further exacerbated by the advantages of establishing a 'Toronto Region' which includes the important outer, but closely linked communities of Barrie, Peterborough, Guelph, Waterloo, Brantford, and Niagara. This urban agglomeration of more than 31,562 km², with 7.5 million people in the GTHA and another 2.8 million in the outer ring [1], was first called the Golden Horseshoe by Westinghouse Electric Corporation president Herbert Rogge in a speech to the Hamilton Chamber of Commerce on January 12, 1954:

"Hamilton in 50 years will be the forward cleat in a 'golden horseshoe' of industrial development from Oshawa to the Niagara River" [2].

Westinghouse established its first Canadian branch in Hamilton in 1897. The company played a major role in developing AC electricity supply from Niagara Falls; at its peak in 1955, there were 11,000 employees at the Hamilton office alone. The company also established a large nuclear manufacturing center in Peterborough. In 2023, Cameco and Brookfield, both Canadian companies, acquired Westinghouse Electric Company, with a key focus on developing eVinci micronuclear reactors and their large-scale AP1000 reactor. Westinghouse recently announced a new engineering design center in Waterloo, with suggestions of up to 8000 new design jobs [3]. Westinghouse provides an excellent example of the integrated nature of manufacturing in the Golden Horseshoe, or Toronto Region.

There are more than 400 companies in each of the energy systems (nuclear, renewables, and grid supply) and automotive/mobility sectors in Ontario; about 80% of those are in the Toronto Region. Despite being home to most of Ontario's manufacturing and post-secondary education and contributing more than 70% of Ontario's GDP (almost 30% of Canada's), the Toronto Region is one of the world's most loosely governed and managed urban areas. The sum of the whole is much less than the competing parts.

Toronto Region (comprised of the CMAs of Toronto, Oshawa, Hamilton, Peterborough, St. Catharines, Kitchener, Brantford, Guelph, and Barrie) has more than 34 transit agencies, 17 electricity distributors, 25 school boards, eight health networks, 25 publicly funded colleges and universities with more than 40 campuses, along with 21 upper- and separate-tier municipalities, and 89 lower-tier municipalities.

The challenge of governing Toronto Region is illustrated with the complexities of obtaining standard, regularly reported data for the region. Statistics Canada provides key information through census metropolitan areas (CMA, see Sec 2.10). A CMA has a minimum population of 100,000 and is largely defined by the commuting practices and predominant location of work of residents. The GTA encompasses almost all of Statistics Canada's Toronto CMA (Toronto CMA includes the City of Toronto and most of the four contiguous regions of Durham, Halton, Peel, and York). However, Burlington, a city within Halton Region, is included in the Hamilton CMA, and the eastern part of Durham Region is split off into a separate Oshawa CMA. The Toronto CMA also includes the northern, fast-growing communities of Orangeville, Essa, New Tecumseth, and Bradford West Gwillimbury in Simcoe County. There is no common, publicly available data set, consistent with political boundaries in the Toronto Region. There is also no government or agency assigned to advocate for the collective best interests of the urban agglomeration.

In 2023, the City of Toronto was 46% of the total GTA population—this will likely drop to below 40% by 2046 (Table 3.1). The four regional chairs of Durham, Halton, Peel, and York (the other half of the GTA) together represent a larger population than the mayor of Toronto. Toronto's share further declines with the addition of Hamilton (to around 40% in 2023). The City of Toronto is less than 30% of the population of Toronto Region (the Golden Horseshoe).

Today the Toronto CMA alone has a population of about 7.5 million, which is 2.5 million greater than Newfoundland, Prince Edward Island, Nova Scotia, New Brunswick, Manitoba, Saskatchewan, and all three northern territories combined.

If the Toronto Region, CMAs were to become a separate province, it alone would have Canada's largest share of population and GDP. A share that continues to grow relative to the country overall.

The City of Toronto population is expected to grow a robust 35% by 2050. This impressive growth is however less than 10% of the growth expected in neighboring regions (area code 905) and the outer ring of Toronto Region. Between 2023 and 2050, the City of Toronto's relative share in Toronto Region will decline from a third to a quarter of the total population. These population figures are fluid, and somewhat arbitrary, as many essential workers and students live outside the City of Toronto, while regularly traveling into the city.[1] For example, more than 80% of City of Toronto police officers live outside the city. Almost half of a police officer's time in Toronto Region is spent in their vehicle, not counting commuting time [4].

Within the 2004, *Places to Grow* report (revised 2005) by the provincial Ministry of Public Infrastructure Renewal, the term *Greater Golden Horseshoe* was introduced and the boundaries extended to include Waterloo Region, Barrie and Simcoe County, Peterborough, and the counties of Brant, Haldimand, and Northumberland [5]. The Greater Golden Horseshoe (Toronto Region) was officially designated under the Places to Grow Act that included a 15% non-resident speculation tax on residential properties [6].

3.1 Why so Many Cooks?

The City of Toronto is represented by 26 elected officials (arbitrarily decreased by the province from 48 during the 2018 municipal election); Durham Region and its eight local municipalities—91 electoral seats; Halton Region and its four local municipalities—67 electoral seats; Peel Region and its three local municipalities—65 electoral seats (this was uncertain from 2022 to 2024 as the province announced dissolution of the Region, and then rescinded the directive), and York Region and its nine local municipalities—99 electoral seats. The communities in the outer ring of Toronto Region include another 650 municipal political representatives in county, city, township, and regional governments (nine county wardens and regional chairs; 68 mayors; approximately 590 councilors).

---

[1] Toronto is also differentiated by area codes, with 416 the inner core (the 6ix) and 905 the regional suburbs. This is only notional as the 416-code is over-subscribed, requiring new area codes. There is also 519 covering much of the west part of the agglomeration, 613 in the east, and 705 in the north, including cottage country.

Table 3.1 Population of the greater golden horseshoe (Toronto region)

	2023	2046
Region of Durham	764,147	1,072,471
Region of York	1,248,989	1,670,264
City of Toronto	3,135,243	4,196,455
Region of Peel	1,689,045	2,600,799
Region of Halton	640,420	1,000,470
City of Hamilton	611,083	860,366
Total GTHA	7,448,507	11,400,825
Northumberland County	93,065	116,677
Peterborough (City & County)	154,424	195,392
Kawartha Lakes	84,726	112,502
Simcoe (including Barrie & Orillia)	583,151	845,876
Dufferin County	71,441	111,897
Wellington (including Guelph)	257,498	395,380
Region of Waterloo	674,489	1,062,209
Brant (including Brantford)	164,182	227,622
Haldimand	128,636	173,058
Region of Niagara	510,226	688,521
Total Outer Ring	2,721,838	3,929,134
Total Toronto Region	10,210,345	15,329,959

Population estimates from Ontario Ministry of Finance, (https://data.ontario.ca/dataset/population-projections), updated July 2023.

The 106 municipal (local) governments of the Toronto Region include: Region of Durham (with Ajax, Brock, Clarington, Oshawa, Pickering, Scugog, Uxbridge, Whitby); Region of York (with Aurora, East Gwillimbury, Georgina, King, Markham, Newmarket, Richmond Hill, Vaughan, Stouffville); City of Toronto; Region of Peel (with Caledon, Brampton, Mississauga); Region of Halton (with Burlington, Halton Hills, Milton, Oakville); City of Hamilton. [inner ring—30 governments]. Northumberland (with Alnwick/Haldimand, Brighton, Cobourg, Cramahe, Hamilton, Port Hope, Trent Hills); City of Peterborough; Peterborough County (with Asphodel-Norwood, Cavan-Monaghan, Douro-Dummer, Havelock-Belmont-Methuen, North Kawartha, Otonabee-South Monaghan, Selwyn, Trent Lakes); City of Kawartha Lakes; City of Barrie; City of Orillia; Simcoe County (with Adjala-Tosorontio, Bradford West Gwillimbury, Clearview, Collingwood, Essa, Innisfil, Midland, New Tecumseth, Oro-Medonte, Penetanguishene, Ramara, Severn, Tay, Springwater, Tiny, Wasaga Beach); Dufferin County (with Amaranth, East Garafraxa, Grand Valley, Melancthon, Mono, Mulmur, Orangeville, Shelburne); City of Guelph; County of Wellington (with Centre Wellington, Erin, Guelph-Eramosa, Mapleton, Minto, Puslinch, Wellington North); Region of Waterloo (with Cambridge, Kitchener, Waterloo); City of Brantford; County of Brant; Haldimand County; Region of Niagara (with Fort Erie, Grimsby, Lincoln, Niagara Falls, Niagara-on-the-Lake, Pelham, Port Colborne, St. Catharines, Thorold, Wainfleet, Welland, West Lincoln); Six Nations of the Grand River [outer ring—76 governments].

3.1 Why so Many Cooks?

In the four regions of the GTA (Durham, Halton, Peel, York), the 322 municipal council seats have considerable overlap. The area is also supported by 28 Chief Administrative Officers, and similar numbers of chief planners, heads of public works, public websites, and municipal headquarters. Each local government has its own Economic Development office with several staff whose performance is measured on progress within their own municipality, often at the expense of a neighboring community.

Municipal officials, elected and appointed, oversee about 25% of the area's total public expenditures and service delivery mandate. Other key public agencies include those administered by the province (about 30% in areas such as education, health, transportation, services such as driver's licenses and social support), federal government services (about 15%), regulated corporations, e.g., telecom and Internet, and government-owned corporations, e.g., LDCs and electricity generators (about 30%).

Natural monopolies such as electricity, cell phones, natural gas, and Internet access are regulated by provincial and federal agencies (with significant cross-subsidization from denser urban communities to more rural areas).

Despite having 348 municipal ridings, the GTA remains under-represented in the provincial legislature and parliament of Canada.

The Province of Ontario is naturally interested in Toronto. In 2020, Toronto CMA represented 52% of the province's economy. The Greater Toronto Area is almost 60% of the province's economy (with a similar share of population) but is represented by only 53 of the 124 legislative seats.

The federal government is similarly interested in Toronto as the area that provides almost a third of the country's economy and remains the country's main port to global connectedness. However, illustrating an even larger bias for more rural and less populated areas, the faster-growing GTA, with almost 20% of Canada's population, is represented by 55 seats in the 338-seat national parliament.

The approximately 650 elected municipal representatives in the Toronto Region are mostly elected to represent a ward, a smaller part of their larger municipality. Their political success is predicated on differentiating themselves among their neighbors. Most people vote in municipal elections based on immediate interests, such as property taxes or re-zoning requests for a proposed building. The challenge extends to the 110 mayors, reeves, and regional chairs in the Toronto Region. Each local leader must promote and differentiate their community, Oshawa versus Whitby, or Mississauga versus Brampton, or City of Toronto and Toronto versus the regions that surround it. Each argues for unique status within the broader urban agglomeration. The political process forces the focus to be on the bifurcated parts. There is no mechanism to account for cross-border priorities. In the Toronto Region, the problem is acute as there is not even a collective agreement at any level of government on what the *whole* is, while the parts grapple with increasingly broad-based challenges.

3.2 The Limits of Boundaries

For forms of Government let fools contest; whate'er is best administered is best.

— Alexander Pope

The need for good governance and the size of the Toronto Region's economic contribution and population (both growing relative to Ontario and Canada) has brought about tensions for more than 100 years. Calls for streamlining local government to better control infrastructure and creating a "Greater Toronto" were made as early as 1907 by William Findlay Maclean, a member of provincial parliament and founder of *The Toronto World* newspaper [7]. This was followed up in 1924 when Ontario cabinet minister George S. Henry proposed a 'metropolitan district' with its own council [8]. Provincial government chose not to act on these recommendations, responding to the opposition from surrounding municipalities.

From 1950 until 1953, the Ontario Municipal Board, under the chairmanship of Lorne Cumming, deliberated changes to local government structure. Opinions ranged from full amalgamation, favored by people like Fred Gardner (Forest Hill Reeve and elected to the Toronto and Suburban Planning Board in 1949), and continued full autonomy, favored by local municipalities. On January 20, 1953, the Cumming's report proposed a compromise solution: a two-tiered government. The province acted on this recommendation and in 1954 established Metropolitan Toronto. This was reorganized in 1967 when seven villages were incorporated into the five boroughs. In 1998 Metro Toronto and the boroughs became the single-tier City of Toronto. In 1971, the neighboring Region of York was established, followed by the Regions of Durham, Halton, and Peel in 1974.

The GTA Task Force, chaired by Anne Golden, was established in April 1995 by Ontario Premier Bob Rae to "provide direction for the future governance of the GTA, including the potential restructuring of the responsibilities and practices of municipal and provincial governments." The final report was presented to then Premier Mike Harris in January 1996. The Report's main recommendations included: equalization and a common assessment of property taxes; establishment of a regional government; disentangling and streamlining local and provincial governance (and service delivery); and a more sustainable urban form, with more efficient infrastructure. The report argued that potential savings from the resulting coordination of services would provide annual savings between $1 billion and $1.5 billion [9]. With the change in provincial government, very few of these recommendations were enacted.

An important development under the Liberal (majority) provincial government and the Smart Growth Secretariat was the 2005 Greenbelt Plan and the 2006 Growth Plan for the Greater Golden Horseshoe. The Greenbelt Plan urged denser and better-connected urban development, setting aside 810,000 ha of protected Greenbelt lands in the Toronto region, in addition to lands protecting the Niagara Escarpment and Oak Ridges Moraine (headwaters of key aquifers).

One of the challenges in sustaining the Greenbelt Plan is pressure by land developers and municipalities to target designated lands for development. Large shares of municipal revenues in fast growth areas like Clarington come from land development charges, and new building construction is one of the largest sources of economic development in most Toronto region communities. The 2023 "Greenbelt Scandal" highlighted the challenges associated with pressures to develop protected lands. In 2022, the majority Conservative Government of Premier Ford removed 15 sites totaling about 2995 ha from the Greenbelt. This provided a minimum increase in land values of $8.28 billion. The value for one area alone in agricultural preserve lands in Durham Region would rise by more than $6.63 billion [10].

Despite the clear imperative, the potential restructuring of responsibilities and practices of municipal and provincial governments in the Toronto Region has been sporadic and delayed. Political vagaries, local government and community opposition to governance changes, plus enormous financial pressures to maintain the status quo and not shift to denser, better-connected development have typically been more than enough to overcome the best of intentions.

Partly in recognition of the problems associated with the governance of urban services in the Toronto Region, and lack of substantive change, some have called for more dramatic changes like establishment of a new province [11] and Charter City Toronto [12]. Sancton (2008) in *The Limits of Boundaries* provides a comprehensive assessment of why a separate political jurisdiction for Toronto region is unlikely, and if it were established, may bring as many new challenges as it hopes to address [13].

The GTA is the minimum urban area needing better coordination. The more strategic area requiring coordination is the Golden Horseshoe, or Toronto Region. The City of Toronto would be less than half the GTA population, and about a third of the Toronto Region. Internal political tensions are likely—cooperation among the regional municipalities is not a historic strength. For example, within the City of Toronto suburbs alone a strong populist leaning of 'us' versus the 'downtown elites' was evident in Rob Ford's election as Mayor in 2010.

Creation of a new province for Toronto Region would likely fracture Ontario into four provinces as eastern, western, and northern Ontario would no longer be contiguous (by population, the smallest of those three potential provinces, Northern Ontario, would likely still have a population larger than PEI, NL, or NB). Services such as electricity supply, waste disposal, and highways would be difficult to partition along new boundaries. Interprovincial trade barriers are already one of the leading causes of Canada's declining productivity (see Chap. 6), exacerbating these barriers would be contested by many.

Probably the most serious opposition to re-drawing government borders would be the (justifiable) concern that this would further erode Canada's overall economic output, rather than work to enhance innovation and productivity (see Chap. 6). A strong opposition from 'concerned Canadians' would likely foment. Rather, a more cooperative, conciliatory form of governance and management is needed in the region.

3.3 Transportation in the Toronto Region—A Spoiled Broth

Toronto's transportation sector is likely where the most damage was done by too many cooks. In *The Lost Subways of North America,* Berman (2023) provides a comprehensive assessment of historical progress in improving transit for 23 North American cities, including Montreal, Toronto, and Vancouver. Toronto's relative lack of progress is glaring and referred to as dysfunctional "political football" [14]. Examples are plentiful. Changes in provincial governments stopped and started several transit lines. Often after extensive design work, and in one case in 1995, the new government under Premier Harris, filled in the partially completed Eglinton tunnel. This one transit line alone has more than a 45-page Wikipedia entry outlining political changes and impacts. The line was first identified in the 1985 Network 2011 transit report. Work began in 1994. Part of the line is now expected to open 30 years later in 2025, after extensive time and cost overruns. In addition to political changes associated with the line, there are significant issues on coordination between the provincial agency Metrolinx, and subsidiary Crosslinx constructing the line, and the TTC, the agency that will operate the line.

The Downtown Relief Line (DRL) is another political football that changed direction several times. Need for the line to address the over-crowded Yonge Line 1 was identified more than 50 years ago. Premier Bill Davis (Conservative) supported the new line as outlined in Network 2011 (1985). The DRL was however shelved when Premier David Peterson (Liberal) was elected. After five more years of planning, the line was again canceled when Premier Bob Rae (NDP) was elected in 1990. Metrolinx re-instated planning for the DRL in 2008; however the line was delayed with Mayor Miller's election in 2010 and his emphasis on "Transit City". Mayor Rob Ford, elected in 2010, re-started transit with a focus on the suburbs, disdainful of "the elitist downtown latte-sipping media and socialist hordes," and arguing "the downtown people have enough subways already." [14]

Andy Byford the TTC Chair at the time again called for the DRL in 2012 and in 2016 a post-Rob Ford Toronto Council voted once again for the DRL. This however was again put on hold when Rob Ford's brother Doug became Premier and canceled the DRL in 2019 (about 20% of design work was completed). A new Ontario Line was proposed. Construction of the new line began in 2021 with opening scheduled after 2031.

Politically imposed starts, stops, and starts again, also plagued Hamilton's LRT, now under construction again. The plan was agreed to in 2015 and land acquisition and design began. The project was canceled in 2019 by the provincial government due to purported cost escalations. Work began again in 2021 with federal government support [15].

Acrimonious debate also arose with the Hurontario LRT proposed to connect Mississauga and Brampton with part of the route traversing Main Street in downtown Brampton. Previous Premier Bill Davis, a long-time Brampton resident, lived on Main Street and opposed the route. Premier Davis was also the Chair of the Mayor's blue-ribbon committee to attract a university to the city. Mayor Jeffrey, who was wooed by Premier Davis to run for mayor, felt blindsided by the former premier's opposition to the much-needed LRT. Ironically, Bill Davis was once chosen North America's Transit Man of the Year back in the 1970s, for killing the Spadina Expressway (with strong support from Jane Jacobs) and using the provincially owned Urban Transportation Development Corporation (UTDC) to build the Scarborough line. Premier Davis passed away August 8, 2021. Debate is still underway on routing for Brampton's Main Street LRT extension.

Toronto's transit challenges are not only delays in new lines. There are examples of poorly planned and constructed lines as well. The five-station Sheppard Line 4 subway aggressively promoted by Mayor Mel Lastman, opened in 2002. The line has operated well below projections, with some stations serving as few as 4269 riders per day (compared to more than 100,000 at stops like Yonge-Bloor—TTC 2022 ridership survey). A bus rapid transit line would have been far more effective at a fraction of the cost. Similarly, the 6.4 km, 6-station Scarborough Line 3, opened in 1985 (closed 2024). The line, built by the Ontario crown corporation UTDC at the province's insistence, was a smaller-scale track using fully automated cars. The Line was plagued with operational challenges, including proprietary (hard to replace) technology, and difficulties operating in inclement weather. Ironically, the Intermediate Capacity Transit System (ICTS), technology developed by the Province of Ontario was highly successful in Vancouver's SkyTrain. The Canada Line is North America's most efficient commuter rail (system availability of 99.9%, punctuality of 99.8%) [16]. The technology, now owned by Alstom, is operating, or planned in more than 10 cities around the world, including New York (JFK Airport), Beijing, Riyadh, Seoul, Detroit, and Kuala Lumpur [14].

Within the Toronto Region, stop-and-start political pressures are evident in other areas such as congestion pricing and the sell-off of Highway 407. In 2017, Premier Wynne (Liberal) arbitrarily disallowed the City of Toronto to impose congestion charging on the Don Valley Highway and Gardiner Expressway (after the city was earlier given authority to do so). In a highly controversial deal, Premier Mike Harris sold Highway 407 with a 99-year lease in 1999. The road is operated to maximize tolls, not to contribute to reduced area congestion. Revenues for the toll road were $1.3 billion in 2022. Premier Doug Ford (Conservative also) agreed in 2022 that selling Highway 407 was a mistake, saying "I would have never sold it." [17] In the same year however, in a populist effort, Premier Ford exacerbated traffic congestion in the Toronto Region by removing vehicle licensing fees (and removing about $1 billion per year in revenues and increasing the provincial deficit).

Traffic congestion in Toronto is among the world's worst. Toronto commuters spend on average, 98 h annually in rush hour traffic and have the longest average travel time in North America [18]. The 2023 TomTom congestion ranking highlights how severe Toronto's traffic congestion is relative to peer cities. In 2023, Toronto had the third-worst traffic congestion in the world, after only London and Dublin. Toronto's congestion was the worst in North America, by a large margin. Mexico City was the next worst city in North America (but 13th worst in the world). New York City, the third worst in North America was 20th in the world. Global rankings (out of 357 cities) for other North American cities: Vancouver—32; Washington DC—55; San Francisco—71; Winnipeg—93; Boston—97; Montreal—103; Chicago—128. Overall travel time in Toronto (24/7) is an extremely poor 29 min per 10 km (compared to the 93-city North American average, including 12 Canadian cities, of 12.5 min). The average speed at 'rush hour' in Toronto was only 18 km/h (North American average 43.7 km/h).

Toronto's congestion costs are estimated at over $11 billion annually, in productivity losses alone [19]. Based on McKinsey [20] and ReThinkX [21] reports, Ontario Tech University drafted a preliminary study *Improved Transportation in the Toronto Region* (2018) investigating the potential of shifting to 'seamless mobility' in the region. The findings corroborate the Board of Trade's estimated congestion costs, and suggest that through integrated mobility (e.g., better transit and ride-sharing) an annual savings of $45 billion is possible (a yearly savings of about $8000 per household) [4]. Some 25 million tons of reduced CO_2 emissions are also possible (at a much lower cost than switching to electric vehicles—which will not address congestion challenges).

Informal discussion with elected politicians around Toronto Region highlights a growing concern. Many believe that financial support for transit infrastructure (and transportation policies such as removing tolls) is now often driven by political support to specific ridings and local policies, e.g., removing tolls or proposing new highways. Transportation in the Toronto region is a complex problem that has increased dramatically relative to urban comparators. The reasons are myriad. Changing political directions, and local public opposition are two important drivers. So too, however, provincial and national (and corporate) support for automobile manufacturing. The desire of many residents for larger homes, with spacious yards is also a key component, and limited provision of alternatives. Undoubtedly though, better transit and less congestion is possible with political and public support. This however requires far more cooperation and sustained coordination.

The growing role of Toronto's Pearson International Airport is another key aspect to consider in the city's transportation sector. In 2022, Canada's total air passengers included 20,880,296 Transborder (USA), plus another 25,031,067 'Other International' [22]. The share of this air traffic by city was: Halifax—< 1%; Montreal—22%; Ottawa—< 1%; Toronto—45%; Winnipeg—< 1%; Calgary—8%; Edmonton—1%; Vancouver—18%.

Global air passenger traffic is expected to surpass pre-pandemic levels and reach 9.7 billion by 2025, with a doubling by 2042 and a 2.5-fold increase by 2052 [23]. Despite Canada's increasing air passenger volumes, the global share is declining as most future travel is expected in emerging economies. In 2023, Canada was the 15th highest ranked country for international air travel. In 2042 and 2052, Canada is not expected to be in the top 20. A more globally focused Toronto Region is a key driver of more international air travel to Canada.

3.4 The Taylor Swift Eras Tour—A Sour Note for Toronto

Many people in Toronto were excited when Taylor Swift announced that her Eras Tour would include six shows in Toronto. The 20-country, 52-city, 152-show tour generated about $3 billion in overall economic activity (hotels, accommodations, travel, and ticket sales). How a major event such as the Eras Tour has so little direct financial benefit for Toronto (or Vancouver) compared to most global cities highlights the funding challenges faced by municipal governments in Canada, especially in the Toronto Region.

The six Toronto shows at the Rogers Center included 290,000 guests. Total revenues for the event were roughly: ticket sales—$110 Mn (~$74 Mn initial sales; ~ $36 Mn resales); hotel accommodations—$20 Mn; food and beverage—$3 Mn; merchandise—$30 Mn, and travel—$68 Mn (taxi, transit, personal vehicle, airfare). Total tax revenues from these expenditures are $40 Mn. The City of Toronto's total share of these tax revenues was about $3.1 Mn (8.6%). The City of Toronto also had police and emergency management staffing overtime costs of about $2 Mn.

Compare this to the $65 Mn total taxes raised for the 3-night concert in Chicago, where the city received $21 Mn in tax revenues (36%). The city-state of Singapore is illustrative as well. The Ministry of Finance supported an exclusivity agreement with the Eras Tour to be the sole venue in the region. More than 264,000 tourists are believed to have traveled to Singapore for the concert providing a $225 Mn economic boost, and about $40.5 Mn in additional tax revenue. Of the 53 cities included in the Taylor Swift Eras tour, Toronto was the only municipal government that likely lost revenue serving as a tour venue. Compare this to the federal government's receipt of at least $15 Mn, and the province of Ontario's $25 Mn in sales tax [24].

3.5 Trends Affecting Toronto

At least three major trends are impacting the Toronto Region more intensely than the rest of Canada: share of foreign-born population and rates of immigration; post-COVID-19 travel and work patterns; and a steady decline in productivity and waning economic growth.

Rising Levels of Foreign-Born Populations

In 2021, 57% of Toronto CMA residents belonged to a visible minority group, up from 51% in 2016 and 14% in 1981. According to the 2021 Census, 47% of Toronto CMA were foreign-born. More than half of Canada's new immigrants settle in the Toronto Region. With such an ethnically diverse community and high rates of new immigrants, the Toronto Region experiences relatively higher rates of ethnic tensions (for Canada) and foreign influence in political campaigns.

Hate crimes more than doubled in Canada between 2009 and 2022 (from 1250 police reported incidents to 3360) [25]. Toronto Police reported a 47% increase in hate crimes from 2022 to 2023, with 2024 likely to see even higher rates [26]. Almost half of the cases are antisemitic in origin; Islamophobic is another large share. Protests and occupations, e.g., University of Toronto, in response to the Hamas-Israel war are increasingly common in Canadian cities, however due to populations and settlement practices of recent immigrants, these events are more common in larger cities like Montreal and Toronto.

Based on 2017 Statistics Canada data, when the total amount of remittances sent from Canada to other countries was $5.2 billion, the Toronto Region in 2024 likely supplies at least a third of total remittances, or some $2 to $5 billion per year. Similarly, in 2023 Ontario hosted 52% of Canada's foreign students, most residing in the Toronto Region (a population increase of about 400,000).

Canada is unique in the relative political power of recent immigrants. Toronto Region's suburban 905 vote-rich area code, with almost four million residents and 30-some federal seats, provides a powerful illustration. For example, Canada is home to the largest national Sikh proportion in the world (2.1%; India is 1.7%) and has the second-largest total Sikh population in the world, after India. Based on the 2021 Census, Brampton, ON has the largest Sikh population (163,260) of any Canadian city; followed by Surrey, BC (154,415), Calgary (49,465), and Edmonton (41,385) [27].

The Town of Markham in York Region has the highest proportion (78%) of visible minorities in Canada, with 45% of Markham's population being of Chinese origin [28]. Any would-be political party or individual aspirant, vying for local, provincial, or federal office in the 905 area, needs to appeal directly to ethnic populations. The influence of foreign governments and independent agents on Canadian democracy is of growing concern, especially at the riding level with opaque nomination races [29].

Changes to Work and Travel Patterns Post-COVID-19

A July 2023 policy brief by Aksoy et al., provided a summary of working from home (WFH) rates in 34 countries around the world. Canada had the highest rate at 1.7 days (April–June 2023 survey) [30]. Canada also had one of the highest rates of WFH desired by employees (2.5 days per week) and planned by employers (1.8 days). This change has a dramatic impact on office vacancy rates and transit patterns in cities like Ottawa-Gatineau (highest rates of government employees) as well as the downtowns of Montreal, Toronto, and Vancouver.

The City of Toronto especially is passing through a period of uncertainty with respect to the future of work and cites. A detailed analysis by the Globe and Mail assessed vacancy rates in 47 commercial buildings (about half) in Toronto's financial district (bounded by John, Queen, and Jarvis Streets and the waterfront). The 47 buildings have about 465,000 square meters of available space, some with more than 50% vacancy (14 with more than 20%). Post-COVID-19 vacancy rates remain high and are still rising, with a noticeable shift to quality Class A buildings (newer, with more facilities, closer to transit). Building valuations may eventually reflect rising vacancy rates, impacting institutional investors as many of downtown Toronto's commercial buildings are owned by institutions such as pension funds and banks [31].

3.5 Trends Affecting Toronto

City of Toronto property tax revenues would also be impacted by changes to building valuations.

As of late 2024, the return of foot traffic to downtown Toronto appears stalled at about 60% of pre-COVID levels (Wednesday is the highest day with 70% pre-COVID levels, and Friday the lowest with about 37%) [31]. The TTC morning and afternoon ridership peaks remain about 30% lower than 2019, while differences midday are only about 11%. GO Transit ridership (train and bus from regions outside city core) remains highly impacted with peak hours barely half of pre-COVID levels (6–9 am and 3–7 pm) affected the most (M-48%; T-63%; W-61%; R-58%, F-41%). Off-peak hours during the week are 78%-86% of pre-COVID rates, while weekends are almost 120% of pre-COVID levels [32].

Brail and Vinodrai's report *Remote Work: Urban Panacea or Curse?* focuses on Toronto and highlights the extent to which remote and hybrid work are highly variable in terms of geography, industry, occupation, education level, and various other socio-economic and demographic characteristics. Negative consequences are mostly realized by women, youth, racialized groups, newcomers, and immigrants, as well as those with less formal education. There are also concerns about the impact of remote and hybrid work on innovation. Toronto needs to reimagine its downtown to increase attractiveness to residents, workers, businesses, and visitors [33]. There is also a need to reassess municipal finance tools as live, work, and play in downtown Toronto shifts more to live and play, with work being done elsewhere.

The higher desire to WFH by Canadians is likely influenced by high congestion levels. Toronto being the most congested city in all of North America (and 3rd in the world—see above) will be particularly challenged to see a wholesale return to offices (and schools) if alternatives are available. Post-COVID-19 travel patterns and work habits are permanently changed. How cities react will be critical. This is especially the case for Toronto Region.

Post-COVID experience in Paris provides warnings for Toronto. Paris also saw massive shifts in office vacancy rates through COVID; however, Q1 2024 vacancy rates were just 2% lower in the downtown core, compared to the La Defense area's vacancy rate over 15% (despite rents being about half those in the downtown core) [34]. The preference for quality urbanism likely extends beyond individual buildings and may be limited in Toronto's core by congestion, declining transit service, and public amenities.

Toronto's Contribution to Canada's Productivity

As Chap. 5 illustrates Toronto is one of the world's most globalized cities (ranked 11th in 2010, 12th in 2020), and Toronto's economic contribution is increasing relative to the rest of Canada. Productivity growth in Toronto would benefit all Canada; however, this needs to be achieved within a carbon constrained environment, i.e., zero carbon by 2050, a much more circular economy (far less resource intensive), an aging society, and higher rates of newcomers.

As we confront wicked problems and complex challenges this century, with a changing climate, and emerging technologies like artificial intelligence, Toronto

Region can drive Canada and the world's future. Similarly, Iqaluit, Montreal, Lagos, Chicago, Paris, and Jakarta will drive Toronto's future.

Enhancing productivity can be achieved through reduced congestion, fewer trade barriers, less friction in supplying urban services. Productivity will also grow from connecting people and communities to the fabric of Canada, as well as maintaining robust connections overseas.

In *Arrival City,* Doug Saunders introduces readers to 30-year-old Maryam Formuli, who immigrated to Thorncliffe with her mother, her brother, his wife and their three children. Maryam hopes to learn a trade or start a business, and soon be joined by her husband. In the meantime, she has no desire to leave Thorncliffe Park:

"When we are in Thorncliffe, we feel like we are in Pakistan or Afghanistan, but when we go downtown, we are in Canada." [35]

Communities like Thorncliffe Park and cities like Brampton (73% visible minorities) and Mississauga (52% immigrants) [36] provide fertile areas where newcomers can put down roots, while also keeping an eye on where they might wish to move next or have their children move out and start their own lives elsewhere. Community networks are critical: social, mobility, and urban services.

Canada's productivity decline is particularly acute the last few years when measured as per capita GDP. Population is growing far faster than productivity growth. In 2023, Canada welcomed 471,550 new permanent residents—around 35% settled in Toronto Region). Around WWI at least a third of Toronto's housing was in unplanned suburbs (informal self-built poorly serviced) [37]. These areas were largely annexed by the city, serviced and formalized.

In the 1920s, new immigrants passing through arrival cities and unplanned edge-communities, often benefited from increasing land values (as they typically owned the land). This opportunity is much less available today: home ownership is far less attainable for immigrants and domestic youth. Toronto Region can still however increase productivity by enhancing the local quality of life. Reducing congestion and increasing mobility, ensuring basic urban services (e.g., litter, drainage), social services such as libraries and basic health provision. These initiatives need to be launched and coordinated by the 106 local governments across the Toronto Region.

The Toronto Region is expected to double in population (more than 20 million people) before it starts to decline after 2100. With projected rates of immigration and migration to the region, this may well be the most globally connected community in the world. Several key industry sectors could develop. These include materials and mining; artificial intelligence, data systems, and security; food and beverage; post-secondary education and training; entertainment; financial services; advanced manufacturing. Toronto Region also can develop a 'work and visit' visa program that encourages shorter-term residencies (perhaps under two years).

The next 50 years will likely see a shift back to more informal living arrangements, e.g., in the 1920s when up to a third of householders were not related (versus less than 10% today). Home ownership pathways, experienced over the last 50 years, are also likely to change, with more rentals (this will be a key resilience and adaptation strategy).

Just as there is a Maslow's hierarchy of needs for humans, there is a hierarchy of sustainability for cities [38]. The basic need of sustainable cities is civility. Civility, which is based on trust and kindness, is the most important tool available when addressing wicked problems associated with cities. This base of civility facilitates peace order and good government, on which productivity is centered.

In the end, only kindness matters.

— Jewel

Asking for more kindness when managing a city of 20 million seems trite, even if it is framed as a key component of civility within the tenets of civilization. Yet, if looking to enhance productivity of Toronto Region, and the rest of Canada, good government, i.e., good management, is imperative. Waste needs to be collected, traffic needs to flow, energy, food, and water need to be readily available, and community systems need to be sufficiently robust to accommodate the numbers of people and climate envisaged.

3.6 The World Needs More Toronto

In 2022, the Golden Horseshoe, or Toronto Region, became the world's 33rd megacity [39] with a population of more than 10 million.[2] There was no notice when the region passed this milestone. The closest announcement was in 2020 when Ontario's Ministry of Finance projected that the Greater Toronto Area (GTA alone) would surpass 10 million population in 2046 [40]. The GTHA (Greater Toronto and Hamilton Area) is expected to reach a population of 10 million by 2038.

By the end of this century, 84 cities are expected to be or have been megacities (about 20 will have surpassed but then declined below the 10 million level). Even if Canada's population surpasses 100 million, Toronto (GTA) is expected to be the only city in the country to reach megacity status. The City of Toronto (as currently configured) is never expected to reach 10 million population, and the City of Toronto alone will also likely always be less than half the total population of the GTA.

There is no automatic level of power or influence attributed to a city when it reaches megacity status. Urban studies also require a relatively consistent definition of a city area or metropolitan region. Commuter-sheds and economic hinterlands can be transient and at-times arbitrary, and claiming to be a larger city, or region, than is credibly warranted may be counterproductive. However, the Toronto Region (i.e., Golden Horseshoe), or GTHA as a minimum, is for all intents and purposes a single urban entity. Benefits would accrue in thinking and acting like a megacity.

Many urbanists and champions for enhanced productivity of the Toronto area encourage a broad analysis. Arguing for stronger cities in Canada, Iveson and Eidelman (2023) urge a *metropolitan mindset*. Their report suggests separate metro areas of Statistics Canada's defined CMAs for Toronto CMA (pop. 6,202,225),

---

[2] Based on Statistics Canada 2021 census; also see Table 3.1.

Kitchener-Waterloo CMA (pop. 575,847) and Hamilton CMA (pop. 785,184). The Oshawa CMA (pop. 415,311) is not included in the Toronto CMA, even though it is a contiguous component of the GTA and is part of the Region of Durham.

Defining the boundaries of metro Toronto are challenging as there are several overlapping local governments; for example, Oshawa CMA is fully in Durham Region; however, most of Durham Region is included in the Toronto CMA. Burlington is included in the Hamilton CMA, although the city is in Halton Region, which is also mostly included in Toronto CMA. Therefore, as a minimum, and reflecting local and regional governments, the Toronto metro mindset should include the combined CMAs of Toronto, Hamilton, and Oshawa. Arguments may arise quickly that the metro Toronto area extends beyond these three CMAs. A powerful example of this is the case of the Niagara Region (contiguous and south of Hamilton CMA). In 2019, GO Transit announced an expansion of daily, regular train service along Lakeshore West to Niagara Falls. Housing price increases followed immediately and were the fastest rising in Canada [41]. With increased connectivity to Toronto core (where many residents are employed) the Niagara Region is projected to double in population by 2041 [42].

The Toronto Region Board of Trade (2022) issued *Think Like a Region* where the economic benefit of a regional approach is also highlighted. A key focus was the technology corridor (also called the Innovation Corridor) from Waterloo to Oshawa. The example of how regional municipalities came together in developing proposals for the Amazon's HQ2 bid is highlighted. The Innovation Corridor comprises the Census Metropolitan Areas of Oshawa, Toronto, Hamilton, Guelph, and Kitchener-Waterloo. The Toronto Region as defined by the Board of Trade does not include the Niagara Region, or other outer ring municipalities [43].

For simplicity, and consistency with global practices and understanding, "Toronto" should include the Toronto, Hamilton, and Oshawa CMAs (as defined by Statistics Canada). This will include the communities of Orangeville, New Tecumseh, and Essa (in Simcoe County) that are not in the area commonly referred to as the GTHA. This Toronto would be considered the inner ring of Toronto Region (aka the Golden Horseshoe) and includes about two-thirds the population of the urban agglomeration, with the outer ring making up another third of the overall population, and including the communities of Kitchener-Waterloo, Niagara, Peterborough, Guelph, Barrie, and Orillia (see Table 3.1).

Iveson and Eidelman (2023) and the Toronto Board of Trade (2022) encourage a metro mindset or 'thinking like a region'. Agglomeration economics would urge as large a community as practicable [44, 45] however congestion, energy systems, and distribution networks force a focus on efficiencies, subsidiarity and where practicable decentralization [46].

As outlined in Chap. 2, cities benefit from scaling. Typically, as a city increases in population, innovation, and economy (productivity) scale superlinearly (at about 1.15), while infrastructure costs scale sublinearly (at about 0.85). From an economic and productivity perspective, bigger is better, but only if negative externalities like congestion, pollution, and safety and resilience considerations are addressed.

3.6 The World Needs More Toronto

A rudimentary calculation for the difference of Toronto Region growing as a connected single urban agglomeration versus a region with municipalities acting more like stand-alone jurisdictions is illustrative.

In 2023, municipal per-person expenditure in Ontario was about $4000. With an estimated 3% rate of inflation, municipal expenditures would likely rise to $7,880 per person in 2046, or $120.8 billion with the population of 15,329,959. This is the minimum baseline cost increase and does not include contingencies for climate adaptation requirements. The cost also does not include provincial expenditures such as transportation, health care, and education.

In 2023, the per-person GDP in Canada was $53,247. Under the current OECD forecast, this would rise to $64,429 in 2046 (the lowest rate of increase of all OECD-members) [47]. This would provide the Toronto Region a combined GDP of $987.7 billion.

If during the next 20 years, efforts are made to minimize intermunicipal barriers and operate more as a single city (urban agglomeration), significant cost savings, and enhanced GDP are possible. Applying scaling experience in other cities suggests a possible sublinear exponent of 0.85 for activities like municipal services and a superlinear exponent of 1.15 for increasing GDP.

Applied to the Toronto Region in 2046 with 15.3 million people, this provides a potential annual savings of $900 million in municipal expenditures and an additional $2.6 billion in GDP. This suggests a minimum economic boost of $3 billion–$4 billion per year if the Toronto Region behaved cooperatively as a common community.

As outlined in Chap. 4, the Great Acceleration of the last 100 years (the First Half) will shift significantly in the next 100 years, the Second Half. Global population will likely peak around 2086. Canada's current fertility rate of 1.3 births per woman [48] is well below the 2.1 births needed for a stable population. The future size of Canada's cities is fully dependent on the ability and willingness of people to move to them, and continued receptivity of Canadians.

The Toronto Region requires collaborative and dynamic governance as the region undergoes major changes over the next several decades. Key objectives include enhancing economic productivity and efficiency of service delivery; increasing the resilience of urban services and the supply of energy and food; moving toward a zero-carbon and circular economy; serving as a key portal into and out of Canada; welcoming some 250,000–300,000 new Canadians per year; anchoring almost half of Canada's post-secondary research and education; serving as a key node in the Great Lakes Region (with Montreal and US counterpart cities such as Chicago, Buffalo, Cleveland and Detroit), and; constructively asserting its aspirations within the Canadian federation and international partnerships. As Chap. 2 outlines, there is a hierarchy of actions for intervention in urban systems. The most powerful means to intervene is to change the narrative, the paradigm of behavior. Working collaboratively as a common urban system is the most powerful way to enhance overall sustainability of Toronto Region.

3.7 Toronto Region to Toronto Global?

One of the most important questions for Canada over the next 40 years is the degree to which Toronto Region emerges as the world's global city. There is an opportunity: positive attributes to achieve this include higher climate resilience (relative to other cities); supportive immigration practices and receptivity; a diverse economy; general adherence to liberal ideals and a rules-based world order; and a solid foundation of basic urban services.

Toronto (through multiple spokespersons) needs to discuss the city's future with current residents, including those in the rest of Ontario and Canada. How open to Toronto Global is the community? How willing are local governments and institutions to work together?

The global mindset requires moving past the more traditional, zero-sum logic that pervades politics, where city leaders and institutions compete for scarce resources [49]. Rather than fighting over pieces of the pie, Toronto Global can increase the size of the pie. However, this is not without cost. Both costs and benefits need to be shared.

Fundamental questions need to be raised and referred to on a regular basis. There will always be national, and provincial, pressure to spread Toronto's potential growth to other areas. A balance is needed between Toronto-specific initiatives to encourage and welcome new residents, with those more broadly for the rest of Ontario and Canada. A growing Toronto Global must be as attractive for Mississauga as it is for Thunder Bay or Winnipeg.

Canadians, and those people living in Toronto Global, will need to articulate and track their progress in the transition toward sustainability. This is a slightly different dialogue than say, how much of the oil sands should be developed, critical minerals mined, or hydro potential captured. Tools exist to mediate most of these thorny issues. The Toronto Global dialogue is more about identifying potential costs and benefits from possible population growth (including more transient and hybrid residents), seeing this as a dynamic opportunity (dependent on many outside influences), and ensuring relative harmony with regions outside Toronto Region.

> **Box 3.1 Rolling Up the Sidewalk: Toronto's Response to Google**
>
> In 2017, Sidewalk Labs, a subsidiary of Google's parent company, Alphabet Inc., signed a Plan Development Agreement with Waterfront Toronto. Together, Sidewalk Labs and Waterfront Toronto (jointly owned by the City of Toronto, Province of Ontario and Government of Canada) would create a Master Innovation and Development Plan for Quayside, a 4.9 Ha area on Toronto's eastern waterfront. Less than three years later, Sidewalk Labs pulled out of the agreement, leaving many to wonder how cities, technology, civics, data, security, capitalism, and genuine well-being might mix at a neighborhood level [50].

The partners believed the area was an optimum site to test new urban infrastructure and integrated data management systems. Sidewalk Labs provided an extensive 1,500-page initial Master Innovation and Development Plan for a smart city to address urban challenges through innovative technologies and new urban design. Public consultations were extensive, verging on too much discussion, while details were still being worked-out. Opposition to the concept centered around concerns over Big Tech [51] (e.g., Facebook and the Cambridge Analytica scandal of 2018), data security, and the role of the private sector in managing urban services.

Jim Balsillie the former Co-CEO of Blackberry wrote in a fearful op-ed, "Sidewalk Toronto is not a smart city. It is a colonizing experiment in surveillance capitalism attempting to bulldoze important urban civic, and political issues." [52] People like Jim Balsillie, Ann Cavoukian the former Privacy Commissioner of Ontario, NGOs, and city officials, were concerned with Sidewalk Lab's insatiable appetite for data and desires on the entire 325 Ha Eastern Waterfront area.

Data systems can be divided into two broad categories within urban areas: (i) opt-in options for citizens, e.g., a community supported ridesharing app, and (ii) public interest solutions, e.g., dynamic traffic signaling and congestion pricing. These were conflated by Sidewalk Labs and within the understanding and concerns of residents. Another key challenge opponents raised was establishment and management of data trusts [53]. Some argued that Toronto had already lost its groove [54] and the reluctance to innovate was another example of the community's insular nature and waning business dynamism [55]. Parochialism and caution were said to have trumped innovation and new approaches to urban service delivery [56].

Looking back as the data and dust settle[3] Sidewalks Labs attempts and failures illustrate a few issues for cities and citizens. Concerns on data security and Big Tech involvement in urban services are warranted, but ironically, even greater liberties are taken through voluntarily provided data from personal cell phones and Internet involvement.

Probably the greatest lost opportunity of Sidewalk Labs was the potential for local government (and their designated institutions) to emerge as a key champion, and overseer, of the public good on data management and security issues. Network effects require data platforms to emerge, and many of tomorrow's key services will be facilitated through these platforms. To date, most of these emerged through the private sector, and Big Tech. For example, platforms like Facebook, Uber, Airbnb, Amazon, Google, and Waze, require access to enormous levels of freely provided personal data. In a world with

---

[3] For example the 2024 debut of the play "Sidewalk Dreams".

shifting geopolitics, reliance on these platforms can be precarious, as illustrated through concerns over TikTok.

Municipalities often play catch-up in regulating these services or wait for provincial and national governments to do so, rather than work collectively to develop similar platforms with the common good as the key driver, as opposed to profit. Cities need to innovate and develop methods to use technology to help solve public interest problems and improve service delivery. These efforts will proceed more quickly and efficiently with the support of credible partners.

In the next decade, local governments will be forced to develop (or try to stifle) platforms for urban service delivery. Innovative business models and management/ownership structures need to emerge. Household data collection and sharing, ride-sharing services, room sharing, neighborhood services, decentralized energy services, and emergency response support, all need to be developed as common goods. All need well-functioning data platforms. Living laboratories such as the defunct Sidewalk Labs in Toronto and the fledgling Oshawa Teaching City [57] are needed. But more important, as highlighted by Sidewalk Labs, these initiatives need a durable (local) public partnership based on the key component of civilization, trust.

References

1. Wikipedia. Accessed 2024–07–02
2. Fast Facts from Hamilton's Past. Retrieved 2024–06–06
3. Westinghouse opens new nuclear engineering hub in Kitchener | CBC News accessed 2024–06–12
4. Ontario Tech University (2018) Improved Transportation in the Toronto Region
5. Government of Ontario (2005) Places to Grow
6. Government of Ontario (2017) Places to Grow Act, Ontario Regulation 416/05
7. Solomon L (2007) Toronto sprawls: a history. University of Toronto Press
8. Colton TJ (1980) Big daddy: Frederick G. University of Toronto Press, Gardiner and the Building of Metropolitan Toronto
9. Greater Toronto Area Task Force & Golden A (1996) Greater Toronto: background reports to the GTA task force. Queen's Printer
10. Office of the Auditor General of Ontario. (2023). "Special Report on Changes to the Greenbelt"
11. Broadbent A (2010) Urban nation: why we need to give power back to the cities to make Canada strong. HarperCollins Canada
12. https://www.chartercitytoronto.ca/
13. Sancton A (2008) The limits of boundaries: why city-regions cannot be self-governing. McGill-Queen's Press
14. Berman J (2023) The lost subways of North America: a cartographic guide to the past, present, and what might have been. University of Chicago Press
15. https://www.cbc.ca/news/canada/toronto/transit-investment-toronto-hamilton-1.6021839
16. Engineering Matters, episode 185 (October 2022) Building Canada Line's Digital Twin
17. https://www.thestar.com/politics/provincial/i-would-have-never-sold-it-doug-ford-says-privatizing-highway-407-was-a-mistake/article_bd312173-45b3-5b3b-bfb1-32f451c69096.html

References

18. Traffic Index ranking | TomTom Traffic Index (2024), Accessed 2024–06–10
19. Toronto Region Board of Trade (2024), Congestion Task Force
20. McKinsey and Bloomberg New Energy Finance (2016) An integrated perspective on the future of mobility
21. RethinkX (2017) Rethinking Transportation 2020–2030. Arbib and Seba
22. Statistics Canada Table: 23–10–0253–01 (2023–07–28) Travel volumes about two-thirds pre-Covid levels
23. Airports Council International (ACI), annual World Airport Traffic Forecasts (WATF) 2024
24. Hoornweg D et al (2025) In preparation
25. Wang JH, Moreau G (2020) Police reported hate crime in Canada (2020) Juristat. Accessed at: https://www150.statcan.gc.ca/n1/pub/85-002-x/2022001/article/00005-eng.htm; Statistics Canada (2023) Police-reported hate crime, by most serious violation, Canada (select police services)," Accessed at: https://www150.statcan.gc.ca/t1/tbl1/en/tv.action?pid=3510006701&cubeTimeFrame.startYear=2015&cubeTimeFrame.endYear=2022&referencePeriods=20150101%2C20220101
26. https://www.cp24.com/news/187-alleged-hate-crimes-reported-in-toronto-in-2024-almost-half-antisemitic-police-1.6908546
27. Census Profile (2021) Census of Population, Statistics Canada
28. Province of Ontario (2026) Ethnic origin and visible minorities survey
29. New York Times, July 21, 2024. A critical gap in Canadian democracy? 'Yawn,' say Canadian Politicians
30. Aksoy CG, Barrero JM, Bloom N, Davis SJ, Dolls M, Zarate P (2023) Working from home around the globe: 2023 Report
31. Globe and Mail, May 4, 2024
32. Toronto Star, What Happened to Rush Hour? March 30, 2024
33. Brail S, Vinodrai T (2024) Remote work: Urban panacea or curse? intergovernmental committee for economic and labour force development (ICE). Canada, Toronto
34. https://brandondonnelly.com/2024/07/07/european-office-vacancy-rates-la-defense-vs-the-paris-cbd/
35. Saunders D (2011) Arrival city: the final migration and our next world. Vintage Canada
36. https://www.ureachtoronto.ca/peel-region/. Accessed 2024–7–27
37. Harris R (1999) Unplanned suburbs: Toronto's American tragedy, 1900 to 1950. JHU Press
38. Hoornweg D (2016) Cities and sustainability: a new approach. Routledge
39. https://worldpopulationreview.com/world-cities
40. https://www.blogto.com/city/2020/02/when-toronto-population-10-million/
41. https://nationalpost.com/life/homes/go-train-fuels-niagara-housing-boom
42. https://niagaracanada.com/go-train-expansion-fuelling-niagara-housing-boom/
43. Global T Like the Toronto Board of Trade, serves to attract global businesses to the region. The organization is funded by the Governments of Canda, Ontario and local municipalities of Toronto, Brampton and Mississauga and Regions of Durham, Halton, Peel, and York.
44. Basu S, Bale CS, Wehnert T, Topp K (2019) A complexity approach to defining urban energy systems. Cities 95:102358
45. Elmqvist T, Andersson E, Frantzeskaki N, McPhearson T, Olsson P, Gaffney O, Folke C (2019) Sustainability and resilience for transformation in the urban century. Nat Sustainabil 2(4):267–273
46. Moore AA, McGregor RM (2021) The representativeness of neighbourhood associations in Toronto and Vancouver. Urban Stud 58(13):2782–2797
47. GDP and spending—Real GDP long-term forecast—OECD Data); Ontario https://www.fao-on.org/en/Blog/Publications/municipal-finances-2020 OECD forecast Canada total GDP 2023, $1,812,901 (USD million), 2046, $2,821,590 (USD million). (Accessed 4–22–24)
48. World Bank, World Development Indicators (2022)
49. Sancton A (2011) Canadian local government: an urban perspective. Oxford University Press
50. O'Kane J (2024) Sideways: the city google couldn't buy. Random House

51. Filion P, Moos M, Sands G (2023) Urban neoliberalism, smart city, and big tech: the aborted sidewalk labs Toronto experiment. J Urban Aff 45(9):1625–1643
52. Globe and Mail (2018) Sidewalk Toronto has only one beneficiary, and it is not Toronto. Opinion (Accessed 12–11–2024)
53. Austin L, Lie D (2021) Data trusts and the governance of smart environments: lessons from the failure of sidewalk labs' urban data trust. Surveill Soc 19(2):255–261
54. Lorinc J (2011) How Toronto lost its groove, and why the rest of Canada should resist the temptation to cheer. Walrus Magazine
55. Guaay F (2024) Opinion: from potential to stagnation—the cost of Canada's insular approach to growth. Financial Post
56. Jacobs K (2022) Toronto wants to kill the smart city forever. MIT Technology Review
57. Oshawa.ca, https://www.oshawa.ca/en/business-development/teachingcity-oshawa.aspx

Open Access This chapter is licensed under the terms of the Creative Commons Attribution-NonCommercial-NoDerivatives 4.0 International License (http://creativecommons.org/licenses/by-nc-nd/4.0/), which permits any noncommercial use, sharing, distribution and reproduction in any medium or format, as long as you give appropriate credit to the original author(s) and the source, provide a link to the Creative Commons license and indicate if you modified the licensed material. You do not have permission under this license to share adapted material derived from this chapter or parts of it.

The images or other third party material in this chapter are included in the chapter's Creative Commons license, unless indicated otherwise in a credit line to the material. If material is not included in the chapter's Creative Commons license and your intended use is not permitted by statutory regulation or exceeds the permitted use, you will need to obtain permission directly from the copyright holder.

Chapter 4
Booms, Busts, and Echoes Around the World: Demographics and the Great Acceleration

In Canada, and the rest of the world, the last 100 years saw two significant trends, growing population and wealth, along with correlated increased environmental degradation.

Global population grew from two billion in the 1920s to more than 7.8 billion in 2020 (with the highest annual growth rate of 2.1% occurring in 1961). Canada became an urban nation in the early 1920s when more than half the population lived in cities. This became the case for the whole world 88 years later in 2008.

The other trend, following from population growth and urbanization, was that wealth, and its associated environmental degradation, grew even faster, especially during the Great Acceleration, beginning around 1950.

Over the last 100 years, no country, per person, has contributed more to the Great Acceleration and the population boom, and possible bust, than Canada. Chapter 4 provides a discussion on the waste and wealth of cities, while this chapter discusses

Canada's impressive track-record of infrastructure building, as it endeavored to accommodate the country's shifting population.

In 1920, global GDP was about $5.8 trillion. By 2020, this had increased more than 20-fold to $126.8 trillion.[1] During this same time, global direct primary energy consumption increased from 17,963 TWh in 1920 to 151,831 TWh in 2020.[2] Municipal solid waste increased from less than 100 million tons in 1920 to more than 2 billion tons in 2020. Atmospheric levels of CO_2 increased from 303 ppm in 1920 to 414 ppm in 2020, well above the 350 ppm level considered 'safe' to avoid large-scale planetary climate changes.[3] Per capita CO_2 emissions peaked globally in 2011; however with growing populations, emissions have increased by more than 35,000 Mt every year since 2014, despite even a 1500 Mt drop in 2020 attributed to COVID-19 [1].

Indicators such as urban populations, carbon dioxide and methane emissions, ocean acidification, forest loss, real GDP, water use, telecommunications, fertilizer consumption, and solid waste generation all follow a similar growth pattern [2]. Levels were relatively stable and low, until around 1950 when they took off on a dramatic increase. The graphs look like a hockey stick on its side [3].

The heart of the Great Acceleration is "scale, scope and pace." [4] The Great Acceleration ushered in a fundamental change in the relationship between humans and Planet Earth where for the first-time humans started to fundamentally affect the biogeophysical cycles and the planet's climate system. Geologist Paul Crutzen brought the term "Anthropocene" into general usage beginning at a conference in 2000. The Anthropocene was proposed to describe a new epoch, recognizing planetary systems shifting away from the Goldilocks-like stable Holocene to a much less steady world. The key driver of the shift to the Anthropocene was the massive increase in energy use.

Fossil fuels, from coal, gas, and oil, along with a rapidly growing urban population, have destabilized critical earth systems. Fossil fuels powered the equipment that moved the earth (humans now annually move 24-times more rock and aggregate than all the world's rivers combined) [5], increased seafood catches, and more than 900 million Ha of the world's forests razed in the last 70 years [6] (about 35 billion tennis courts). With more than 1 billion tons, the combined weight of humans (390 Mt), plus domesticated livestock and pets (630 Mt) far exceeds the planet's remaining wild terrestrial (20 Mt) and marine (40 Mt) biomass [7].

The Great Acceleration follows the path of urbanization, as it is driven mostly by urban residents, who use most of the world's energy and hold most of the wealth. The Great Acceleration will come in three phases. From 1950 to about 2000, the Great Acceleration was largely driven by people living in the cities of OECD-member

---

[1] 2020 GDP values down from 2019 level of 130.7 trillion due to COVID-19. 2021 value increased to $134.8 trillion post-COVID. Expressed in 2017 international $. From Our World in Data (accessed 2024–06,016).

[2] Our World in Data, Global direct primary energy consumption. Accessed 2024–06-16.

[3] The annual rate of increase in CO_2 concentrations first surpassed 1.0 ppm in 1965, 2 ppm in 1979, and in 2023 the concentration increased by 2.82 ppm (reaching 419.3 ppm). Values from NOAA climate.gov.

countries. At the start of this century, the BRICS countries (Brazil, Russia, India, China, and South Africa) took over as the world's largest drivers of wealth and waste. For example, growing quickly, China surpassed the USA as the world's largest emitter of CO_2 in 2006 and in 2022 generated more than twice as much CO_2 (11.4 Gt vs 5.1 Gt) [8]. The third phase of the Great Acceleration has only just started and will not likely be declining before 2050. This phase includes the impacts from the cities of Sub-Sahara Africa, the last, and largest region to urbanize (see Table 2.2).

Building the Bones: Canada's Long-Lived Infrastructure

The Great Acceleration—the massive increase in energy and materials use, waste generation, and wealth—is facilitated through infrastructure. Infrastructure, often unseen, anchors urban systems. Our cities are built on systems supplying electricity, water supply, solid waste and wastewater management, transportation, and telecommunications. These technological systems are physically located in the city or connect cities. They are often under foot or overhead, and unobserved when functioning well. These systems interact with each other, users, systems managers, and governments, through a broad suite of local and global linkages.

Over the last 100 years, Canada's infrastructure, like bones underlying the urban corpus, was built within and between cities. This substantive base however still needs $11 trillion to $22 trillion of additional infrastructure investment between now and 2067, just to maintain current standards of living [9]. In addition to building Canada's cities to accommodate an expected doubling of population, new infrastructure will need to emit far less GHG emissions and be hardened for a less hospitable climate.

Canada's history is inextricably linked to its public infrastructure. The Canadian Pacific Railway (1885), the St. Lawrence Seaway (1959), the Trans-Canada highway (1962), and the Anik communications satellites (1972–1975), connected communities across the country, promoting commerce and spreading settlements. Canada's abundant hydropower and nationally supported nuclear power, provided relatively low-cost electricity that helped power growing cities and drove the country's manufacturing sector.

The Erie Canal completed in 1825 connecting Lake Erie to the Hudson River is an example of strategic infrastructure that catalyzed a regional economy. The Canal shifted the locus of economic concentration from Philadelphia to New York City and opened the American interior to settlement [10].

Federal government support of infrastructure (e.g., roads, railways, canals, ports, pipelines) was often believed necessary to open the hinterland to economic activity, provide producers with access to international markets, support northern development, and knit the country together. Infrastructure such as communications, roads, broadcasting, and postal service encouraged integration of regional economies [11].

Ballantyne Pier in Vancouver provides a powerful example of the contribution of long-lived infrastructure. The pier, completed in 1923, took advantage of the 1914 opening of the Panama Canal, providing an alternative route to Europe for wheat and timber, rather than only eastward by rail. Today the pier continues to support the City's economy by anchoring the West Coast's burgeoning cruise industry.

Montreal's two-kilometer Victoria Bridge was the world's longest when opened in 1859. The bridge remains a vital link today carrying Route 112 and rail connections. Calgary's Centre Street Bridge opened in 1916 and continues to be an integral part of the city's transportation network.

Toronto is also served by long-lived, critical infrastructure. For example, opened in 1947, the 828-km Highway 401 is North America's busiest highway as it passes through Toronto. As an economic artery, the Highway anchors much of the province's manufacturing, and increasingly its residential development patterns. Highway 401, complemented with other routes like Highway 400 and the Queen Elizabeth Way, links Montreal and Quebec City to Windsor, Detroit, and eventually Chicago. The highways serve the agricultural sector while also linking northern Ontario and western Canada to southern Ontario, the USA, as well as Quebec and Atlantic Canada.

Another two key infrastructure components serving Toronto include Adam Beck and Pickering Nuclear Generating Stations. Taking advantage of the enormous potential power of the Niagara River as it falls 99 m between Lakes Erie and Ontario, Adam Beck I, commissioned in 1921, and Adam Beck II, commissioned in 1954, provide almost 2000 MW of reliable, low-carbon electricity. Enough to supply 1.8 million Ontario homes (or 3.6 million German homes).

Adam Beck I was the world's largest infrastructure project when constructed, complete with an onsite hospital. The Adam Beck stations were named after Sir Adam Beck, the former Mayor of London and member of the Ontario legislature, who served as the first chairman of the province's Hydro-Electric Power Commission. Beck was a strong proponent of publicly owned electricity supply, arguing "the gifts of nature are for the public." Hydroelectric development along the Niagara River, and eventually the St. Lawrence River, was made possible through the Canada-US Boundary Waters Treaty of 1909, establishment of the International Joint Commission (IJC) and subsequent water-course specific agreements [12].

Adam Beck tried to introduce a network of publicly owned interurban ('radial') railways with trolleys powered by electricity from Niagara Falls. In the 1919 post-war election, Beck lost his seat to the United Farmers of Ontario (UFO) as they swept the Conservatives out of power [13]. Beck's goals of interurban railways were quickly abandoned by the UFO as they focused on rural services and cheaper electricity. The railways were also quickly losing out to the increasing popularity of automobiles.

Growing alongside relatively cheap electricity, automobiles became a large share of Ontario's manufacturing sector. Electricity demand surpassed the available hydro-capacity in the 1950s. With uncertain coal supplies, Canada's growing nuclear capabilities post-WWII, and a supportive federal government, Ontario decided to pursue nuclear power to augment hydrosupply.

After about five years of design work and successful demonstration reactors, construction began in 1965 at the Pickering Nuclear Generating Station, just east of Toronto. The first unit of the Pickering A facility (four units, total of 2000 MW) began operation in 1971. The first Pickering B unit (four units, total of 2000 MW) began operation in 1983. Discussions are now underway to refurbish Pickering B to extend operations to at least 2060.

There are over 15,000 dams in Canada, many helping the country to be the fourth-largest producer of hydroelectricity in the world (after China, Brazil, and the USA). Quebec alone has 61 hydroelectric dams with a combined capacity of 38,400 MW, almost half the country's total hydrogeneration capacity. Hydropower has such a prominent history in Canada that it is one of the few countries that often uses 'hydro' interchangeably to refer to electricity. Almost all of Canada's infrastructure for electricity generation was led by the provinces. Ontario developed most of its readily available hydrocapacity by the 1960s and subsequently shifted to nuclear and coal generation.

Quebec pursued large-scale hydroelectric development in the twentieth century. British Columbia, Manitoba, and Newfoundland and Labrador also developed extensive hydrocapacities. Manitoba Hydro, for example, dammed the Nelson River, to use Lake Winnipeg as one of the world's largest reservoirs.

Almost half of British Columbia's hydroelectric power is from the Columbia River which was facilitated by the Colombia River Treaty between Canada and the USA. The treaty was ratified in 1964 and led to the construction of four significant dams (combined, over 12,000 MW).

Urban infrastructure in Canada from 1920 to 2020 initially saw much greater support from the federal government; however from 1961 to 2002, the federal share in infrastructure finance dropped from 25 to 7%, while the municipal share doubled from 25% to half [14].

Infrastructure support from the federal government became less reliable, often focused on highly visible investments. For example, ad hoc federal financial support was part of the bids for the Olympics, PanAm and Commonwealth games, and World Expositions such as Montreal 1967 and Vancouver 1986. Large-scale energy infrastructure, with particularly high regional significance, might also receive federal support, e.g., the $35 billion twinning of the Trans Mountain pipeline to transport bitumen from the oil sands to Asian markets, and the more than $20 billion in proposed support for interprovincial high-voltage electricity transmission lines, such as the Atlantic Loop [15].

As outlined in Chap. 1, cities face a key challenge from uncertain financial support for infrastructure and inconsistent policy interventions by the federal and provincial governments. This is most evident in capital-intensive infrastructure sectors such as transit and transportation. The federal government, for example, believed an additional airport on the eastern side of Toronto would be necessary and in 1972, 7,500 Ha of agriculture and future development lands were expropriated. Preliminary construction activity was halted in 1975 when the Government of Ontario declared it would not build the roads or sewers needed to service the site. In the absence of a land-use decision, for more than 50 years this prime real estate has been unavailable for development or granted guaranteed protection.

Mirabel Airport provides another example of infrastructure plans failing to materialize. In 1969, 39,250 Ha were expropriated for Montreal's second international airport. The area, larger in extent than the entire city of Montreal was a compromise location (the Provincial government wanted the site to be more central within Quebec, while the federal government wanted the airport to also service Ottawa and western Montreal hinterland). Construction was fast-tracked, without the planned surface links to Montreal (more than a 50-min drive away), and the airport opened in the fall of 1975, in time to support the 1976 Summer Olympics. Mirabel airport was developed for an estimated 17 million passengers per year. The actual level never exceeded 3 million passengers. The last passenger flight was in 2004, and in 2016 the passenger terminal was demolished.[4]

In 2023, Federation of Canadian Municipalities (FCM) calculated infrastructure costs per dwelling unit in a Canadian city. Each housing unit requires an average investment of $107,000 in municipally owned capital assets.[5] The report referenced a 2021 City of Ottawa study that showed servicing low-density greenfield development cost the city $465 per person per year, while high density infill development provides the city with $606 per person per year of revenue (a difference for the city of more than $1000 per person per year depending on where they live). The differential was supported by a 2023 study that determined infrastructure costs in Vancouver for a single-family house were five to nine times more expensive per person compared to infrastructure services to multiresidential development.

The FCM estimate of $107,000 in municipally owned capital assets does not include the capital costs associated with electricity supply, gas transmission, telecommunications, schools and hospitals, or buildings to support governments and utilities. The share of capital versus operating costs varies by service. For example, solid waste management budgets are typically around 80% operating costs and 20% capital, while water supply can be 80% capital and 20% operating costs.

Canada's infrastructure benefits from networks. Some, like the ability to plug in a new electronic device safely and reliably anywhere in the country, date back to the decision to use AC electricity rather than DC more than 100 years ago. Every home or business connected to roads and transit, electricity, and telephone networks is part of a powerful collective system connected regionally and globally. As more people join the network, service costs drop per connection, and collective benefits increase (see Chap. 2). For example, access to clean water reduces waterborne diseases for everyone, and roads enable emergency vehicles and deliveries to reach all buildings [16]. Canada's relatively mature levels of infrastructure will benefit as global networks grow. They also are well positioned to drive innovation and support the shift to lower carbon and greater climate resilience.

As human bones are made of calcium, the skeleton of cities is made of concrete. Concrete is sand and gravel, glued together by cement. Cement is made by firing lime, clay, and other minerals in a kiln to about 1350 °C. The process is energy intensive, as well as responsible for about 8% of the world's CO_2 emissions.[6] The share of global CO_2 emissions from cement is a quick way to observe global city-building status. In 2022, China generated 48% of the world's CO_2 emissions from concrete,

---

[4] From Wikipedia, Montreal-Mirabel International Airport. Accessed 2024–06-22.

[5] FCM, November 23, 2023. New research: Canada's housing challenge is also an infrastructure challenge.

[6] Nature 597, 593–594 (2021).

India—10%, Vietnam—3.7%, USA—2.6%, Turkey—2.4%, and Canada less than 0.5% [17].

Concrete became transformational in city building with the addition of steel reinforcing. This allowed concrete to be formed into dams, bridges, and buildings. The concrete bones are connected with copper and aluminum transmission wires, steel framing, underground pipes for water and wastewater (concrete, steel, and plastic), timber, glass, bricks and stone for surface facing, aggregate for roads and the base of buildings, elevators, and myriad materials. A typical 50-story building likely weighs 300,000 t[7] and a single-family Canadian house 300 t.[8]

4.1 Boom, Bust, and Echo: Canada's Continued Contribution?

In 1921, Canada's population was 8.8 million. The largest cities were Montreal, Toronto, Winnipeg, Vancouver, and Hamilton (average population 319,371). One-hundred years later, Canada's population grew more than four-times to 37 million, with the largest cities Toronto, Montreal, Calgary, Ottawa, and Edmonton having an average population of 1,578,4878.

Canada's fertility rate was about 5.54 at Confederation in 1867. The value declined steadily as health care, education, and sanitation advanced, while maternal health and infant mortality rates significantly improved. By 1940, the fertility rate had dropped to 2.69. However, as the post-war baby boom emerged this level spiked up again hitting a new peak of 3.88 in 1960. Since then, with the introduction of the contraceptive pill in 1960 and shifts in family dynamics, Canada's fertility rate has steadily declined to less than 1.35, a level lower than the USA and most other countries.[9]

In 1996, the seminal book *Boom, Bust and Echo* David Foot and Daniel Stoffman argued that demographics 'explains about two-thirds of everything' [18]. The book outlined how baby boomers in Canada (those born between 1947 and 1966) would drive much of the country's economy and urban form. Demographics will continue to drive the shape and size of Canada's cities; however, a global demographic assessment is needed when planning for Canada's next 100 years. Canada's prosperity no longer depends solely on what is happening at home, but rather rests on the population booms overseas. These booms will also rapidly shift to busts.

The UN Population Division (2022 update) projects global population to peak at 10.43 billion in 2086. Annual population growth, that hit a peak of 2.3% in 1963

---

[7] Based on 50 floors at 3000 m^2 per floor, with each floor using 1000 m^3 of reinforced concrete (@2.7 t/m^3), with a live load ~ 1500 t, 800 t cladding and miscellaneous (~5000t/floor), plus base slab 50,000t.

[8] Author estimate.

[9] For comparison, national fertility rates: South Korea—0.8; Spain, China—1.2; Japan, **Canada**, Finland—**1.3**; Austria, Russia—1.4; Germany, Netherlands, Sweden—1.5; UK, Brazil, Australia—1.6; USA, Ireland—1.7; France, Mexico—1.8; India—2.0; Philippines—2.7; Egypt—2.9; Pakistan—3.4; Tanzania—4.7; Nigeria—5.1; Niger—6.7; OECD-average 1.6. World Bank 2022.

(0.9% in 2023), is projected to drop below zero before 2090. Projections beyond 2100 are unavailable, however, the trends (with zero population growth) suggest a rapid decline in the first half of the twenty-second century. What the likely stable human population ends up being, post-Great Acceleration and Great Deceleration, is not clear, if in fact the human population stabilizes. Best estimates suggest an eventual stable planetary population of three to six billion by around 2150 [19].

In 2023, Canada's population grew by 3.2%, a level not seen since 1957. However, a big difference of this population growth is that the birth rate in 1957 was 3.8 live births per woman. In 2023, this birth rate was down to 1.35, well below replacement levels of 2.1. Almost all of Canada's population growth is now from immigration (96.7% in 2023).

Canada's new economic growth will be through technology and immigration. The number of potential available immigrants might increase as large swaths of the planet become less hospitable due to a changing climate; however, integrating climate refugees may be different than welcoming those seeking better economic opportunities. Also, Canada's declining birth rate is shared among most high-income countries. The growth of world population is starting to decline and age everywhere, the difference is the rate. Canada will need to compete with other countries to attract and retain immigrants.

4.2 From Immigration to Migration

Canadian cities are among the largest recipients of immigrants in the world. Immigrants make the conscious decision to emigrate for a variety of reasons, but usually immigration signifies a permanent, international re-location. The host country needs to offer the potential immigrant a quality of life sufficiently attractive to uproot their lives. Immigration is predicated on a national mindset and is usually accompanied with a lengthy legal process.

Migration is either internal (within a country) or international and tends to be more reactive, perhaps fleeing war or drought, or excessive heat. Migrants do not necessarily intend to permanently re-settle. All immigrants are migrants, but not all migrants are immigrants.

The stable climate niche that enabled people to establish agriculture, cultures, and countries is changing rapidly. As the stable climate niche shifts, significant migration away from the equator should be anticipated in the next 50 years [20]. Immigration was mostly about benefitting the receiving country. Migration needs to be seen more as a key component of sustainability, perhaps less as an advantage for receiving countries, and more as a benefit to humanity overall.

Gaia Vince in *Nomad century: how climate migration will reshape our world* [21] and Parag Khana in *Move: How Mass Migration Will Reshape the World – and What It Means for You* [22] describe the next phase of human civilization, as one that is both mobile and sustainable, suggesting that disruption is inevitable, but tragedy is not. Migration can potentially contribute to sustainability transitions when it enhances

4.2 From Immigration to Migration

well-being while not exacerbating structural inequalities or environmental burdens [23].

Governance—how people work together to achieve what they cannot achieve on their own—has long been a central focus of both migration and sustainability studies [23]. Mitigating disaster will fall disproportionately to countries like Canada. However, the paramount role played by national governments in the last 100 years (as the number of countries worldwide more than tripled), is likely to shift as regions and cities, and groups of countries, assume new roles in governance.

Butterflies, birds, and fish naturally migrate in response to climate change. Similarly, human migrations reflect the biology of *Homo sapiens* and should be embraced, not feared. The degree that this migration is embraced by a country versus cities and regions within that country may continue to widen. Sonia Shah in *The next great migration: The beauty and terror of life on the move* [24] presents human migration not as a crises but as a natural solution to human resilience. Canadian cities will increasingly be called on to be part of that solution.

Canada plays an active, and growing role in welcoming migrants, international students, and refugees. In 2020, globally there were about 280 million migrants (3.6% of global population). Canada was the eighth most welcoming country for migrants, after the USA, Germany, Saudi Arabia, Russia, UK, France, and United Arab Emirates. The leading sources of Canada's migrants are India (720,083), China (699,190), Philippines (633,547), UK (537,504), and USA (273,226) [25]. The world's top three refugee resettlement countries in 2022 were Canada (44,772), the USA (42,365), and Australia (7773) [25]. In 2023, Canada provided $8.6 billion in international remittances (about 1.3% of the global total of $669 Bn).[10,11] Beginning 2025 however, Canada reduced immigration numbers, reflecting growing public concern with affordability and housing availability.

Canada has a fundamental choice to make. Although the choice is largely being made by others, Canadians will still need to decide how much they fight these inexorable forces, and how much they adapt, and try to maximize potential benefits. Suggesting *Canada* has a fundamental decision is challenging as sometime between 2040 and 2050 Canada's populations is expected to be more than 50% foreign-born [26].

Geography is often presented as destiny, and geography certainly presents Canada with a fundamental choice. Canada is one of the few countries that may still be able to choose how large it wants to be in 2120. The ability to decide on population size however is time-limited as the world moves closer to peak population. Over the next 100 years, differences between immigration and migration will also blur around the world. Temporary foreign workers (and migrants) may have fewer placed-based rights, but possibly greater ability to maintain personal and economic linkages to their country of origin.

---

[10] The cost to send remittances from Canada remains high (6.5%). Higher than Australia, France, Germany, Italy, Korea, Saudi Arabia, UK, or USA, and well-above the G20 target level of 5% and more than twice the SDG target of 3% by 2030 (World Bank, 2023).

[11] World Bank Indicators 2024.

4.3 The Rise of Post-secondary Education

In 1920, Canada had 23,314 students in universities (86% male, 14% female [27]; 43% in Arts and Science, 14% in medicine, 13% in engineering [28]; 32 institutions in 14 cities). In 2020/21, Canada had 1,405,683 students in universities (74 institutions in 48 cities, 57% female, 43% male, 17% international students). Additional college students in 2020/21 included 616,017 Canadian and 142,308 international students in some 148 institutions. Total post-secondary students increased from 0.3% of population in 1920 to 3.8% university and 2% college student enrollment in 2020.[12]

Funding for universities was limited. American foundations were highly influential: From 1920 to 1940, the Carnegie and Rockefeller foundations provided nearly $8 million to Canadian universities, more than 5-times typical provincial funding.[13]

Cities such as Kingston and Guelph, Ontario; Antigonish and Cape Breton, Nova Scotia; and Prince George, British Columbia; have traditionally been influenced by relatively large numbers of post-secondary students living in the community; however cities like Toronto, Vancouver and Montreal, should anticipate growing populations of post-secondary students, especially foreign students. This influences housing and local employment levels. About 5.8% of Canada's total population were post-secondary students in 2020. For some cities, post-secondary students could comprise as much as 10% of the population. Numbers are likely to ebb and flow as issuance of student visas varies; however, long-term trends suggest that foreign students will remain an important component of urban dynamics.

4.4 Booms and Busts in Canadian Cities

As Table 4.1 illustrates the relative share of population in Canadian regions changed markedly in the last 100 years. This is likely to continue to 2120. The population of the Atlantic Region, for example, is expected to increase almost six times from 1920 to 2120; however, the relative share of Canadian population will decline from 11% to less than 6%. The population of Canada in 2120 is difficult to estimate reliably as it is almost entirely based on immigration (and migration) levels that will depend on both the receptivity of Canadians and the interest of potential immigrants. Also, based on the estimated level of non-permanent resident population of 6.5% in 2024, a higher value for non-permanent residents should be estimated for Canadian cities in future, especially the larger cities.

With increasing climate migration and more informal residency patterns, e.g., foreign students and temporary workers (and possibly longer-stay tourists, plus 'golden visas'), Canadian cities should plan for a minimum 10% 'floating population'

---

[12] Statistics Canada 2024.

[13] See Footnote 12.

Table 4.1 Canada population 1921 to 2120

	1921	2021	2120
Canada	8,788,483	36,991,981	105,000,000 (est)
Atlantic Canada	1,000,328 (11.4%)	2,409,874 (6.5%)	6,000,000 (5.7%)
Quebec	2,361,199 (26.9%)	8,501,833 (23%)	20,000,000 (19%)
Ontario	2,933,622 (33.3%)	14,223,942 (38.5%)	41,500,000 (39.5%)
Manitoba	610,118 (6.9%)	1,342,153 (3.6%)	4,000,000 (3.8%)
Saskatchewan	757,510 (8.6%)	1,132,505 (3.1%)	3,500,000 (3.3%)
Alberta	588,454 (6.6%)	4,262,635 (11.5%)	13,500,000 (12.9%)
British Columbia	524,582 (6%)	5,000,879 (13.5%)	14,000,000 (13.3%)
Northern Territories	12,145 (<1%)	118,160 (<1%)	2,500,000 (2.4%)

From Statistics Canada national census for 1921 and 2021 (accessed 3/10/24). Atlantic Canada includes Newfoundland and Labrador in 2021 and 2120. Northern Territories includes NWT, Yukon, and Nunavut in 2021 and 2120.
2120 author's estimate (see Sect. 3.5). Population likely to have peaked before 2120 and, and like most countries, undergoing a slow decline by 2120.
Non-permanent resident (NPR) populations of 1.5 million (2021 estimate) and 10+ million (2120 estimate) not included. NPR share of 2024 population 6.5%.
Based on Statistics Canada high growth scenario for 74 million by 2068 (56.5 million under medium growth scenario), and 55,521,000 in 2043 (with Atlantic Canada—2,675,000; QC—10,196,700; ON—21,147,200; MB—1,944,400; SK—1,696,700; AB—7,158,000; BC—7,360,500; NT, YK, NU—162,600).
Under the medium growth scenario, the UN projects that the global population will peak in 2086, at just over 10.4 billion people (the low growth scenario has population peaking in 2054 at 8.9 billion; the high growth scenario does not see global population peak this century).
Hoornweg and Pope (2014) projected populations under SSP1 scenario to 2100 for Toronto (9,287,353), Montreal (6,242,855), and Vancouver (3,868,103).

of additional residents, and possibly as high as 30%. These higher numbers are particularly important in planning emergency response plans, and potential evacuation and re-location plans.

Typically, municipalities need to plan for infrastructure services and housing based on future population estimates. However, as Table 4.2 outlines, local governments have little ability to produce these estimates through typical actuarial practices. Births and deaths are relatively straightforward to project; however in 2022–2023, these represented less than 1% of the population variations in Montreal, Toronto, or Vancouver. More than 99% of population changes were from immigration, emigration, and migration within and between provinces. Local governments have very limited influence on any of these values. As immigration and other factors of population change, e.g., foreign student permits, remain largely the exclusive purview of the federal and provincial governments, local governments will need to be provided these estimates by senior levels of government, or not be expected to develop official plans based on population projections.

Canada's ability to welcome immigrants has served cities well, but this national bonhomie is not guaranteed. Resentment toward immigrants has always ebbed and

Table 4.2 Changes to populations July 1, 2022, to June 30, 2023

	Montreal CMA	Toronto CMA	Vancouver CMA	Canada
Births	40,888	59,514	22,147	357,903
Deaths	34,486	39,984	18,518	330,379
Births minus deaths	6402	19,530	3629	27,524
Immigrants	49,151	134,526	52,799	468,817
Net emigration	3526	5085	6113	35,337
Emigrants	10,354	21,944	13,350	94,576
Returning emigrants	6828	16,859	7237	59,239
Interprovincial migration	− 8975	− 16,092	− 4795	
Intraprovincial migration	− 20,624	− 93,024	− 18,399	
Non-permanent residents	106,836	181,733	92,529	697,701
Population	4,502,177	6,804,847	2,971,853	40,097,761

The following census metropolitan areas (CMA) had more deaths than births in 2022–23: St. John's, NL; Moncton, Fredericton, NB; Saguenay, Sherbrooke, Trois-Rivieres, Drummondville, QC; Kingston, Belleville-Quinte West, Peterborough, St. Catherines, Windsor, Sudbury, Thunder Bay, ON; Kelowna, Kamloops, Victoria, Nanaimo, BC. Canada overall likely to have more deaths than births around 2030. Of the 111 census agglomerations, only 27 had more births than deaths. Temporary foreign worker approvals in 2022 (222,847) and 2023 (239,646), Employment and Social Development Canada.
Source: Statistics Canada, Table 17–10-0149–01, released 2024–05-22.

flowed, due to actual or perceived shortages of economic opportunities and services, e.g., rising cost of housing, challenges to get a campsite or admission to government programs.

Canada's mosaic requires the ability to knit disparate groups together. Groups can keep their cultures and aspirations, but they need to accommodate other groups with similar aspirations. How communities are woven together into a tapestry rather than having rigid shards tear things apart remains a Canadian challenge.

Perhaps the most challenging trend is how willing new immigrants are to move to Canada, and how amenable these people are to remain in the country and make it home. This will be impacted by global shifts in population and how welcoming our climate is relative to other parts in the world with changing climate patterns.

In the USA in 1964, 77% of Americans trusted the federal government most, or all of the time. Since 2010 this level dropped and stayed below 20% [29]. In Canada, only 33% of citizens believe that most people can be trusted. This is consistent between genders but varies markedly between age cohorts (26% younger vs 39% older), income levels (26% lower vs 40% higher) and by education (lower 25% vs 46% higher) [30]. In the 28 country assessment, Canada and the USA are roughly in the middle (33% believe 'most people can be trusted'), while China India and Indonesia (56%), Netherlands and Saudi Arabia (48%), and Australia (41%) have higher levels of trust; and Russia (24%), Mexico (23%) Japan (21%), and Brazil (11%) have lower levels of trust.

Canada's welcome mat for immigrants and migrants has shifted in size and prominence over the years. In 1923, the Government of Canada revoked the head tax, a large fee charged to Chinese people entering Canada, replacing it with the Chinese Immigration Act, which virtually halted all immigration from China. Over the next 24 years, only 44 Chinese migrants entered the country. By 1931, less than four percent of B.C.'s population was Chinese. In 1914, Canada turned away the Komagata Maru carrying 376 passengers, mostly Sikhs from Punjab, India.[14] In 1939, Jewish refugees fleeing Nazi Germany aboard the MS St. Louis were denied entry into Canada on arbitrary grounds [31].

In the first half of the twentieth century, Canada favored European and American immigrants, and excluded certain groups including Communists, Mennonites, Doukhobors, and nationalities whose countries had fought against Canada during WWI, such as Austrians, Hungarians and Turks [31]. In 1967, Canadian immigration practices changed with the introduction of a points system to rank potential eligibility of applicants. Race, color or nationality had no bearing on points.

How rising levels of immigration might affect levels of trust in Canada will be a key issue for the next several decades. Also, how trust might vary by region or city. For example, 50% of Toronto's population was born in another country, while only 20% of Montreal was: Will that impact future levels of trust? Or, if rural high school dropout rates remain around 16%, while urban rates are 9%, will that affect future levels of trust? Canada's peak year for immigration (by percent of population) was 1913 with the arrival of 400,000 newcomers (5.3% of the population), mostly of European origin. If future immigration approaches these levels again, but with more disparate cultural and ethnic groups, will national trust levels be impacted?

4.5 Canada's Population in 2120

Canada's population in 1921 was 8.8 million and more than quadrupled to 37 million in 2021 (about 1.5% annual growth). As Table 4.1 suggests Canada's population could increase to about 105 million by 2120. Statistics Canada provides population estimates to 2068; with 56.5 million projected under the medium growth scenario. There is considerable uncertainty associated with population projections, both globally and for Canada specifically.

Vollset et al. in *The Lancet* project global peak population of 9.73 billion occurring in 2064, with population estimates in 2100 of 8.79 billion under the reference scenario and 6.29 billion under the SDG scenario (enhanced education and contraception levels) [19]. Specifically for Canada, the article projects peak population occurring in

---

[14] The Canadian Encyclopedia (2024).

2078, with a population of 45.17 million, and estimates for 2100 a population of 44.09 million under the reference scenario (36.8–53.16 range), and 37.08 million under the SDG scenario (32.09–43.28 range). This is markedly different from Table 4.1.

UN Department of Economic and Social Affairs, World Population Prospects 2022, projects global population will peak around 2086 at about 10.4 billion, and then start a slow decline.[15]

In 2023, Canada experienced high population growth, with a 3.2% annual increase to reach 40,769,890 on January 1, 2024.[16] More than 97% of this growth was from immigration (and does not include temporary foreign workers and students). Canada's population in 2120 is therefore conditional on the number of immigrants admitted per year. If the country were to maintain a 1% growth rate (a third of what was experienced in 2023), population in 2120 would be 108 million. A 2% growth rate (two-thirds that experienced in 2023) would lead to a population in 2120 of 290 million. Maintaining 2023 growth rates (which no one is suggesting) would lead to a Canadian population of 769 million in 2120.

Planners in Canadian cities face an extremely difficult challenge, they have almost no control over what populations are likely to be over the next 100 years (Table 4.3). National and provincial governments may be able to continue to welcome immigrants as was the case for the last 100 years, however there are two big differences that need to be considered. Unlike the previous 100 years, Canada's birth rate is among the lowest in the world. Almost all increase in population will need to come from immigration. Second, during the next 100 years the world will very likely reach peak population. Climate change will cause large numbers of migrants and potential immigrants, and Canada's geography is particularly welcoming. However, with most high-income countries aging quickly, Canada may find itself competing for immigrants. Technology developments may also impact Canada's receptivity to immigrants, e.g., robotics and healthcare provision. Overlaying these trends is a growing backlash against immigration, especially in the USA with the re-election of President Trump. Toronto will likely continue to play a key role in Canada, although better coordination is needed (see Chap. 2). The Toronto Region now makes up more than two-thirds of Ontario's population, with more than half the population foreign-born. How the Toronto Region grows with Montreal and the rest of the Great Lakes Region will determine much of Canada's future prosperity. Two other growth centers are particularly important to Canada's future prosperity as well. Calgary-Edmonton will likely be the fastest-growing, and Vancouver, within Cascadia, might anchor the megaregion's growing influence along North America's west coast.

---

[15] See https://population.un.org/wpp/.

[16] Statistics Canada, released 2024-03-27.

Table 4.3 Population of Canada's largest cities 1871–2021

1871	1921	1951	1971	2001	2021
Montreal 107,225	Montreal 618,506	Montreal 1,021,520	Montreal 1,214,351	Toronto 2,481,494	Toronto 2,794,356
Quebec City 59,699	Toronto 521,893	Toronto 675,754	Toronto 712,786	Montreal 1,039,534	Montreal 1,762,949
Toronto 56,092	Winnipeg 179,087	Vancouver 344,843	Edmonton 438,152	Calgary 879,003	Calgary 1,306,784
Halifax 29,582	Vancouver 163,220	Winnipeg 235,710	Vancouver 426,256	Ottawa 774,072	Ottawa 1,017,449
Saint John 28,805	Hamilton 114,151	Hamilton 208,321	Calgary 403,319	Edmonton 666,104	Edmonton 1,010,899
Hamilton 26,716	Ottawa 107,843	Ottawa 202,045	Hamilton 309,173	Winnipeg 619,544	Winnipeg 749,607
Ottawa 21,545	Quebec City 95,193	Quebec City 164,016	Ottawa 302,241	Mississauga 612,000	Mississauga 717,961
London 15,826	Calgary 63,305	Edmonton 159,631	Winnipeg 246,246	Vancouver 545,671	Vancouver 662,248
Portland, NB 12,250	London 60,959	Calgary 129,060	Laval 228,010	Hamilton 490,268	Brampton 656,480
Kingston 12,401	Edmonton 58,821	Windsor 120,040	London 223,222	Surrey 347,825	Hamilton 569,353
Canada 3,485,761	Canada 8,788,483	Canada 14,009,129	Canada 21,888,679	Canada 30,007,094	Canada 36,991,981

From Statistics Canada various census years. Compiled in Wikipedia (accessed 2024–06-13). Populations may vary due to provincial amalgamations of municipalities, e.g., Toronto 1971 to 2001. Canada total, same census year.
Population by Census Metropolitan Area (CMA) in 2021: Toronto 6,202,225; Montreal 5,928,040; Vancouver 2,642,825; Ottawa–Gatineau 1,488,307; Calgary 1,481,806; Edmonton 1,418,118; Quebec City 839,311; Winnipeg 834,678; Hamilton 785,184; Kitchener–Waterloo 575,847.

References

1. World Resources Institute (2024) The History of carbon dioxide emissions
2. Steffen W, Sanderson RA, Tyson PD, Jäger J, Matson PA, Moore III B, Wasson RJ et al (2005) Global change and the earth system: a planet under pressure. Springer Science and Business Media
3. Steffen W, Crutzen PJ, McNeill JR (2007) The Anthropocene: are humans now overwhelming the great forces of nature. Ambio 36(8):614–621
4. https://www.cbc.ca/radio/ideas/great-acceleration-1.6814569#:~:text=American%20environmental%20historian%20John%20McNeill,pace%2C%22%20McNeill%20told%20IDEAS. Accessed 2024–06–18
5. Cooper AH, Brown TJ, Price SJ, Ford JR, Waters CN (2018) Humans are the most significant global geomorphological driving force of the 21st century. Anthropocene Rev 5(3):222–229
6. Our World in Data (2024) Deforestation and forest loss
7. Greenspoon L, Krieger E, Sender R, Rosenberg Y, Bar-On YM, Moran U, Milo R (2023) The global biomass of wild mammals. Proc Natl Acad Sci 120(10):e2204892120

8. Our World in Data (2024) Global CO_2 emissions from fossil fuels
9. Future Cities, Deloitte (2020) Building our urban futures: inside Canada's infrastructure and real; estate needs
10. Bernstein PL (2005) Wedding of the waters: the Erie canal and the making of a great nation. WW Norton and Company
11. Gramlich EM (1994) Infrastructure investment: a review essay. J Econom Literat 32(3):1176–1196
12. International Joint Commission (2016) The boundary waters treaty of 1909
13. Mentzer MS (2006) Irrational optimism in a declining industry: Sir Adam Beck's interurban railway proposal. Manag Organ Hist 1(4):371–384
14. Harchaoui TM, Tarkhani F, Warren P (2004) Public infrastructure in Canada, 1961–2002. Canadian Public Policy, 303–318
15. van de Biezenbos K (2022) Lost in transmission: a constitutional approach to achieving a nationwide net zero electricity system. Osgoode Hall Law J 59(3)
16. Chachra D (2013) How infrastructure works: inside the systems that shape our world. Riverhead
17. Our World in Data (2022) Share of global CO_2 emissions from cement
18. Foot DK Toffman D (1998) Boom, bust and echo 2000: profiting from the demographic shift in the new millennium
19. Vollset SE, Goren E, Yuan CW, Cao J, Smith AE, Hsiao T, Murray CJ (2020) Fertility, mortality, migration, and population scenarios for 195 countries and territories from 2017 to 2100: a forecasting analysis for the Global Burden of Disease Study. The Lancet 396(10258):1285–1306
20. Xu C, Kohler TA, Lenton TM, Svenning JC, Scheffer M (2020) Future of the human climate niche. Proc Natl Acad Sci 117(21):11350–11355
21. Vince G (2022) Nomad century: how climate migration will reshape our world. Flatiron Books
22. Khanna P (2021) Move: how mass migration will reshape the world and what it means for you. Orion Books
23. Adger WN, Fransen S, Safra de Campos R, Clark WC (2024) Migration and sustainable development. Proc Natl Acad Sci 121(3):e2206193121
24. Shah S (2020) The next great migration: the beauty and terror of life on the move. Bloomsbury Publishing USA
25. McAuliffe M, Oucho LA (eds) (2024) World migration report 2024. International Organization for Migration (IOM), Geneva
26. Statistics Canada (2022) Immigrants make up the largest share of the population in over 150 years and continue to shape who we are as Canadians
27. Harris RS (1976) A history of higher education in Canada 1663–1960. University of Toronto Press. http://www.jstor.org/stable/https://doi.org/10.3138/j.ctt1vxmbqp
28. Usher A (2018) Higher education strategy associates
29. Pew Research Center (2024) Public trust in government: 1958–2024
30. Ipsos (2022) Interpersonal Trust across the world
31. The Canadian Encyclopedia (2024)

References

Open Access This chapter is licensed under the terms of the Creative Commons Attribution-NonCommercial-NoDerivatives 4.0 International License (http://creativecommons.org/licenses/by-nc-nd/4.0/), which permits any noncommercial use, sharing, distribution and reproduction in any medium or format, as long as you give appropriate credit to the original author(s) and the source, provide a link to the Creative Commons license and indicate if you modified the licensed material. You do not have permission under this license to share adapted material derived from this chapter or parts of it.

The images or other third party material in this chapter are included in the chapter's Creative Commons license, unless indicated otherwise in a credit line to the material. If material is not included in the chapter's Creative Commons license and your intended use is not permitted by statutory regulation or exceeds the permitted use, you will need to obtain permission directly from the copyright holder.

Chapter 5
The Wealth and Waste of Cities

On July 12, 2019, the *USA Today* newspaper (24/7 Wall Street) published an article headlined *Canada produces the most waste in the world. The US ranks third* [1]. Using data from the World Bank's *What a Waste: A global review of solid waste management* [2], the top ten list of waste-generating countries was as follows [3]: (10) Serbia @ 8.9 t/cap, (9) Ukraine @ 10.6 t/cap, (8) Luxembourg @ 11.8 t/cap, (7) Sweden @ 16.2 t/cap, (6) Armenia @ 16.3 t/cap, (5) Finland @ 16.6 t/cap, (4) Estonia @ 23.5 t/cap, (3) USA @ 25.9 t/cap, (2) Bulgaria @ 26.7 t/cap, and (1) Canada @ 36.1 t/cap.

The numbers and ranking were a surprise to people familiar with waste volumes across countries. The same report referenced, for example, Canada's waste generation rate as 1.94 kg/capita/day (0.71 t/cap or about 2% of the value given in the newspaper article). In the report, Canada's waste volumes are less than those of the USA (2.24 kg/capita/day), although higher than the world average of 0.74 kg/capita/day [3]. However, in the USA Today article, in addition to municipal solid waste (MSW), Canada's waste volumes included from a partially completed annex 181 million tons of agriculture waste, and 1.12 billion tons of industrial waste.[1] The industrial waste was mostly tailings and earth moving from oil and gas operations (especially the oil sands) and mining activities. For the USA Today headline, Canada topping the world's waste generation list was largely a function of data availability.

Cities and countries are ranked every day in newspapers and magazine articles across the world. The world's most livable cities, countries with the highest greenhouse gas (GHG) emissions, best quality of life, most sustainable cities, most innovative country, etc.: these articles often anchor a publication's best-selling issue. As the *USA Today* article illustrates, the rankings can be subjective and often reflect which metrics were used in the assessment.

When assessing waste volumes or GHG emissions, the question of attribution arises. If more than 80% of Canada's oil and gas, or mined materials like gold, and over 95% of manufactured vehicles [4] and potash [5], are exported, how much of the impact should be attributed to Canada, rather than the final customer? This question is answered differently when responding from a country versus a city perspective.

Over the last 50 years, international frameworks emerged for countries to measure emissions and impacts. For example, under the Intergovernmental Panel on Climate Change the 195 member countries provide their total territorial GHG emissions—all emissions that take place within their national territory. Add these 195 territorial inventories, plus a few global sectors like aviation, maritime shipping, and extraterritorial military activities, and you have a relatively reliable estimate of the global total (57.1 Gt CO_2e in 2023 [6]). This inventory will be further refined as emissions from wildfires and melting permafrost are also included.

Challenges with the territorial emissions approach arise when a business or city wants to know its total emissions or impacts across the full lifecycle of all activities. These impacts may take place within several jurisdictions and countries. Unilever, for example, operates in 190 countries [7] and cities like Montreal, Toronto, and Vancouver have material links to every country in the world. Supply chains need to be measured, even if they might travel through several countries. This is especially important for cities, as they tend to be the terminus of multiple supply chains.

Recognizing the need for corporate, and eventually community, down-scaled GHG emissions inventories that could reflect vicarious or embodied emissions (emissions generated on behalf of the entity but done so through a third party or in another jurisdiction), the World Business Council for Sustainable Development (WBSCD) and World Resources Institute (WRI) began to work together in 1997 to develop a standard methodology consistent with national (territorial) GHG inventories. The concept of scopes 1, 2, and 3 was introduced to account for where the emissions

---

[1] To identify the largest producers of waste, USA Today 24/7 Tempo calculated the special waste and regular municipal solid waste per capita produced by each country, using data from the World Bank's "What a Waste" global database, last updated in September of 2018. The report provides estimated values for total municipal solid waste (MSW) as special waste categories of agricultural waste, construction and demolition, e-waste, hazardous waste, industrial waste, and medical waste. Most special waste values are ad hoc, dependent on data availability.

were generated while ensuring globally consistent national, local, and corporate emissions inventories and reducing the challenges of double-counting emissions. In 2006, the International Organization for Standardization (ISO) adopted the WRI-WBCSD Corporate Standard as the basis for ISO 14064-I: Specification with Guidance at the Organization Level for Quantification and Reporting of Greenhouse Gas Emissions and Removals [8].

Customers and investors are keen to know a company's (and presumably a community's) overall GHG emissions. For example, investing in lower emitting companies like Apple, Google, and CIBC emits just 0.05, 0.17, and 0.2 kg respectively of CO_2e per C$100, while high emitting companies like Teck Resources, Barrick Gold, Suncor, and Arcelor Mittal are several orders of magnitude higher at 25.12, 26.1, 35.4, and 745 kg respectively of CO_2e per C$100 invested. Investing in Arcelor Mittal over Apple results in about 15,000 times more carbon per dollar invested [9].

Attributing environmental impacts like GHG emissions and solid waste generation, and how these are shared between countries, companies, and communities, is still in its infancy. However, as more information is collected and regularly published, trends emerge. Canada with higher rates of resource companies listed on major exchanges, such as the TSX, may have even higher overall GHG emissions. How GHG emissions relate to the triple planetary crises (climate change, air pollution, and biodiversity loss) and are integrated within ESG (environmental, social, governance) considerations is critical. Cities (e.g., urban areas and CMAs) may increasingly take on this task as they are best positioned to do this in an apolitical manner that does not need to favor corporate proponents (see Box 5.1, Scope of the Challenge).

Although not easy to answer unequivocally, questions like 'does Canada produce the most waste in the world?' and 'are Canada's cities among the most energy- and materials-intensive in the world?' are important. Like estimating GHG emissions, the answer to questions like these differ from a country or city (community) perspective. Answering the question from a city perspective should include a more comprehensive assessment, like how progressive businesses attempt to account for all activities throughout their supply chain. The concept of Scopes 1, 2, and 3 does not (yet) apply to solid waste and biodiversity, but here too businesses (and communities) will attempt to quantify impacts across the overall lifecycle of corporate and community activities.

The fashion industry concerned with labor practices in upstream factories, or cellphone manufacturers ensuring adequate environmental safeguards throughout the entire supply chain, is driven by customer preferences. So too will cities be asked to help quantify the overall impacts of their businesses and citizens.

Answering the question 'does Canada produce the most waste in the world' requires a territorial response (like GHG emissions—accounting for all waste generated within Canada's sovereign territory). Answering the question 'are Canada's cities among the most energy- and materials-intensive in the world' requires a more comprehensive response that includes Scopes 1, 2, and 3 emissions as well as expected economic growth (Table 5.1).

Table 5.1 Distribution of Canada's GDP in 2020 and estimated 2070

	2020	2070 (estimate)
Toronto	27%	35%
Montreal	11	10
Vancouver	8.3	10
Calgary	4.9	8
Edmonton	4.2	6.5
Ottawa-Gatineau	4.3	6
Rest of Canada	40.3	24.5

City Share of Provincial Economies (by GDP) 2020

St. John's, NL	44%
Halifax, NS	58%
Moncton-Saint John, NB	43%
Montreal, QC	55%
Toronto, ON	69% (Toronto Region), 59% (GTHA), 52% (CMA)
Ottawa-Gatineau, ON&QC	7% (9 and 4%, respectively)
Winnipeg, MB	66%
Regina-Saskatoon, SK	52%
Calgary-Edmonton, AB	66%
Vancouver, BC	60%

2020 values from Statistics Canada. Table 36-10-0468-01 (2023-12-06), Gross domestic product (GDP) at basic prices, by census metropolitan area (CMA). GTHA includes CMAs of Oshawa and Hamilton. Toronto Region also includes CMAs of Peterborough, St. Catharines, Kitchener, Brantford, Guelph, and Barrie. Vancouver also includes Abbotsford CMA. Montreal, Edmonton, Calgary unique CMAs. Ottawa-Gatineau includes ON and QC CMAs. Total Canadian GDP in 2020, $2,083,990 ($\times 1,000,000$)

2070 author estimate based on historic growth trends 1970 to 2020 and population projections to 2070

5.1 The Waste and Wealth of Canada's Cities

Canada's relatively high per capita MSW generation rates were corroborated by the Conference Board of Canada's Municipal Waste Management 2020 report [10]. The study used OECD data to compare 20 countries. Canada had the highest rate of per capita MSW generation—more than double the lowest levels in Japan [11].

In addition to high levels of solid waste generation, Canadians are also among the world's highest primary energy consumers (Table 5.2), about five-times the global average, and per person twice the average of other high-income countries. Canada's corresponding per capita GHG emissions are also among the world's highest.

As outlined in Table 2.1, global wealth grew even faster than increases in urban populations, or energy use, and waste generation (as GDP per person). This is especially the case for Canada. From 1880 to 2020, Canada's economy grew by 1375%

5.1 The Waste and Wealth of Canada's Cities

Table 5.2 Primary energy use per person in 2022

Canada	102,160 kWh
USA	78,754
High-income countries (average of 1.24 Bn people)	56,469
China	31,051
UK	30,098
Upper-middle income countries (average of 2.78 Bn people)	27,390
World average (7.95 Bn people)	21,039
Brazil	17,300
India	7143
Lower-middle income countries (average of 3.19 Bn people)	6658
Low-income countries (average of 704 Mn people)	1231
Ethiopia	872
Democratic Republic of Congo	411
Somalia	217

Canadians use about 471 times more energy per person than someone living in Somalia; about five-times the global average and almost twice as much energy as other high-income countries. Per person greenhouse gas (GHG) emissions are proportional. Add to this world leading rates of solid waste generation (material consumption) and water use, and per person, Canadians lead the world in environmental degradation.
From Our World in Data, 2024. Data source: U.S. Energy Information Administration (2023); Energy Institute - Statistical Review of World Energy (2023); Population from World Bank, 2022. A human's "natural" metabolic rate is ~ 90 watts. As shown above, a Canadian's 'social metabolism' is ~ 11,700 watts, or the equivalent energy requirements of 12 elephants.[2]

(faster than any OECD-member country, including the USA at 1267% and the world average of 897%; Table 5.3).

Cities in Canada consume more than 50,000 tons of food per day. About 65% of that is wasted in shipping, preparation, and spoiling. About 20% is excreted as human waste. The food is packaged in about 10,000 tons of paper, plastic, glass, and difficult-to-recycle multilaminates like drink boxes, coffee cups, and foil pouches.

From an energy and materials perspective, Canada is a post-urban country. Rural residents use 30–40% more energy and materials and generate correspondingly more waste and GHG emissions than their urban counterparts. Canada's more remote communities, where much of the electricity is still generated by diesel fuel, and some consumables are air-freighted, can have annual carbon footprints over 50 tons per year per person.

Canadian cities were front and center in the Great Acceleration, with some of the world's highest per person energy and materials use. Canada's energy and material use is high from both the production (supply) and consumption (demand) sides.

---

[2] Adapted from: [74].

Table 5.3 Per capita GDP in 'Western Offshoots', 1600–2020 (1990 dollars)

	Australia	New Zealand	Canada	United States
1600	400	400	400	400
1700	400	400	430	527
1820	518	400	904	1257
1870	3273	3100	1695	2445
1900	4013	4298	2911	4091
1920	4766	5641	3861	5552
1970	12,024	11,189	12,050	15,030
2000	21,540	16,010	22,198	28,129
2020	27,237	20,773	24,998	33,426

All values in 1990 Geary–Khamis (international) dollar; 1600–2000 OECD (2006), 2020 Our World in Data, accessed April 8, 2024. Percentage growth 1870–2020: Australia 732%; New Zealand 628%; **Canada 1375%**; USA 1267%; Europe 963%; OECD Average 1278%; Latin America 726%; East Asia 1609%; Sub-Saharan Africa 193%; World 897%. https://ourworldindata.org/grapher/gdp-per-capita-prados-de-la-escosura?tab=chart&country=East+Asia+%28AHDI%29~South+Asia+%28AHDI%29~Middle+East+%28AHDI%29~North+Africa+%28AHDI%29~Latin+America+%28AHDI%29~Western+offshoots+%28AHDI%29~Western+Europe+%28AHDI%29~Eastern+Europe+%28AHDI%29~Sub+Saharan+Africa+%28AHDI%29~CAN~AUS~NZL~USA

Canada's cultural DNA is largely as supplier of energy and resources, hewers of wood, and drawers of water. This started with trade in beaver pelts for the hats of London's rich and timbers for navy ships. However, with a mindset of resource supply and ample land area, Canada's cities became some of the world's most energy- and resource-intensive. Canadians generate more solid waste per person than almost any other country. So too, energy consumption. A resident of Montreal uses more than twice the electricity of a resident in Berlin or Stockholm. Canada's vehicle fleet is the most fuel inefficient in the world [12].

The Great Acceleration in Canadian cities is still accelerating, but not as quickly, and there is a sense of collective bracing for impact as the imperative to decelerate emerges. Most scientists, engineers, and urban practitioners recognize that the current energy and material demands of cities are at levels that cannot be sustained, leave alone increased to meet the world's 3 billion new urban, and more affluent, residents expected over the next 50 years. The issue is not shortage of supply. Material to be mined and processed is available. However, biophysical systems, such as the global atmosphere, nitrogen pollution, and biodiversity loss, are beyond safe earth system operating limits. The planet cannot adequately assimilate the impacts of this material supply while also maintaining critical biosystems.

Oil and gas, wheat, asbestos, mining (about half the world's mining companies are listed on the Toronto Stock Exchange [13]) anchored much of Canada's economy. Hundreds of resource towns emerged across the country, e.g., Kitimat, BC; Fort McMurray, AB; Uranium City, SK; Flin Flon, MB; Sudbury, ON; Murdochville, QC; Glace Bay, NS; and Labrador City, NL.

5.1 The Waste and Wealth of Canada's Cities

The United Nations Environment Program (UNEP) regularly estimates national levels of material consumption (and production) [14]. The Global Resource Outlook provides compelling evidence on the interlinked nature of material use with climate change, pollution and biodiversity loss. A related UN publication tracks progress on the Sustainable Development Goals (SDGs, e.g., 12.2.2 Domestic Material Consumption per capita - All Materials) and is particularly noteworthy for Canada. In 2022 (latest year available), Canada's per capita material consumption was a world-high 52.91 tons per person [15] (more than five-times the global average).

A related UNEP publication by the International Resource Panel, *The Weight of Cities: Resource Requirements of Future Cities* is also relevant to Canadian cities [16]. The report divides the world into 19 geographic regions with Canada and the USA comprising *Northern America*. Northern America is a consistent outlier compared to the rest of the world. Northern America's urban density is the world's lowest (less than 1500 person/km^2, compared to global average of about 7000 p/km^2, or a high of 11,500 person/km^2 in Southern Asia). Northern America is also an outlier with the highest rates of urban domestic material consumption and by far the highest urban final energy consumption (355 GJ/person, compared to global average of about 40 GJ/person; 2010 values).

The UNEP International Resource Panel provides summary data sets for each of the world's G20 member countries. Again, Canada is an outlier for the high volume of energy and materials used. The Natural Resource Use in the Group of 20, Status, Trends and Solutions, Canada summary sheet provides the following information [17]:

- Domestic extraction remained stable at 37 tons per capita (G20 average was 15 tons/capita in 2015).
- The extraction and processing of natural resources accounted for more than 40% of Canada's total climate change impacts from both a production and consumption perspective (the G20 average was approximately 50% from both perspectives *combined,* i.e., half of Canada's total).
- Resource extraction and processing caused almost 40% of outdoor particulate matter-related health impacts.
- Material-related climate change impacts in Canada were mainly caused by the extraction and refinery of petroleum, the extraction of natural gas, cattle farming, and mining of chemical and fertilizer minerals.
- Climate change impacts remained much higher (about double) than the G20 average.
- Materials with large climate impacts (petroleum, natural gas, and beef) were mostly consumed by households, especially for mobility, heating, and food.
- The construction and motor vehicle manufacturing sectors were the largest industrial users of climate-intensive materials.
- From a production perspective, land-use-related biodiversity loss was lower than the G20 average and was mainly caused by forestry activities. However, from a consumption perspective, land-use-related biodiversity loss was comparable to

the G20 average due to imports of beef, oil seeds, vegetables, fruits, and nuts from regions with high ecological value (i.e., outside Canada).
- Canada has been a net exporter of biomass and fossil resources and a net importer of minerals since 2006.
- Food imports caused higher water stress impacts in the countries of origin than biomass exports from Canada (mainly wood).
- For all material types, net value added was higher inside Canada than outside.
- Material footprint and all environmental impacts per capita remained higher than the G20 average. Reducing the consumption of impactful resources like petroleum (particularly for mobility) and beef could help lower these impacts. Furthermore, material-related impacts could be reduced with the design of material-efficient infrastructure and fossil fuels (natural gas) by constructing energy-efficient buildings.

Few Canadians would think Canada has been a net importer of minerals since 2006. Canada's per capita material consumption is a world-high 52.91 tons per person. This is in addition to the much higher than per person GHG emissions and one of the world's highest levels of primary energy use (about 5-times the global average). Most of this profligate energy and material consumption is driven by the shape of Canadian cities and the lifestyles of Canadians. Per person, Canadians produce and use more materials and energy than any other country. A Canadian has a greater impact on earth systems than any other nationality (see Sect. 5.8, Blame Canada, Rich Canadians, or all the Rich?).

5.2 What a Waste

Consume or die. That's the mandate of the culture. And it all ends up in the dump.
—Don DeLillo, Underworld

Estimating waste generation rates is challenging. Solid waste management is usually a municipal responsibility although in most high-income countries more than half the waste, especially in the industrial, commercial, and institutional (ICI) sector is contracted by individual generators to the private sector, making data collection difficult. Final disposal volumes are also difficult to compile as waste haulers are not always required to report collection and disposal volumes.

Municipal solid waste (MSW) is all waste material generated within a community and is usually divided into residential (about 45% of the total) and ICI (about 55%) waste streams.

Environment and Climate Change Canada reports that in 2016, 10.2 million tons of waste generated from the residential sector, 11.5 million tons from the industrial, commercial, and institutional (ICI) sector, and 3.2 million tons from the demolition, land-clearing, and construction (DLC) sector were disposed as follows:

- 20.3 million tons disposed in landfills in Canada;

5.2 What a Waste

- 3.8 million tons exported to the USA; and
- 0.85 million tons incinerated—primarily to produce energy [18].

Canada's solid waste challenge is especially acute in remote communities. Shipping from these communities is prohibitively expensive, and waste from the communities requires local dumpsites, often with limited environmental safeguards.

Iqaluit provides a useful example. The community had a dump fire the summer of 2014 that caused significant nuisances, including school closures and measurable air-quality impact [19]. The fire burned for months.

As discussed in Sect. 5.1, MSW is only a small (< 5%) fraction of total material consumption. For example, the world's most consumed resources are water (4000 km^3), sand (50 Gt), aggregate (gravel), topsoil (36 Gt), fossil fuels (11 Gt), and cement (4.1 Gt) [20].

Values like zero waste and zero carbon, even if watered-down with "net", are approached asymptotically; the last few percent requires Herculean efforts to achieve. Absolutes also open the door for obfuscation, bitter arguments, and far-too-long delays.

Cities, the cradles of civilization, will need to play to their strengths in the approach to net zero and arbitrate with pragmatism, empathy, equity, and increasingly important, quickly. Cities, the urban areas that consume the most and generate much of the world's wealth and culture, will drive the narrative of the twenty-first century. In Canada, this will fall disproportionately to the five large urban regions (Vancouver, Calgary-Edmonton, Toronto, Ottawa-Gatineau, and Montreal), plus critically important regional communities in the north, the Maritimes, and rural resource towns.

Canada is a loose federation, with strong provincial powers. Canada's provinces often act as down-scaled nations. And yet Canada's cities are among the world's best for livability. These assessments however do not include the relative impact on the planet's ecosystems or current economic contribution compared to potential ability. Here Canada's cities fare among the world's worst.

Canadians consume far more energy than global comparators. In 2021, Canadians consumed 400 GJ of energy per person, compared to a global average of 80 GJ, USA of 310 GJ, Europe 132 GJ, China 116 GJ, India 21 GJ, and Africa with a paltry 11 GJ per person. This resource-intensive lifestyle is also apparent in solid waste generation rates and GHG emissions. Canadians again are among the world's leaders in consumption and waste.

Person-for-person Canadians have a roughly fivefold greater impact on planetary systems than the global average; more than 35-times a resident of Africa. The social metabolism of one Canadian is equivalent to more than 12 elephants (Table 5.2). The excuse that Canada is a large, cold country, that is dark for much of the winter carries little weight as more than 80% of citizens live in cities that could be significantly less energy- and material-intensive, and comparator communities where energy use is less than half of Canada's are common.

5.3 A Visitor's Perspective

Tourist visits to Canada are relatively low compared to international comparators. In 2021, Canada was the 27th most popular tourist destination (93rd based on tourists per resident) [21]. Toronto, Montreal, and Vancouver are among the world's top cities by population size, and Toronto (12th), Montreal (41st), and Vancouver (70th) are among the most globalized cities (Table 5.4); however, Toronto was only the world's 53rd most visited city in 2023, Vancouver the 85th (Montreal was not in the top 100) [22].

Considering that Toronto is now more than half foreign-born, visitor numbers suggest the city, and country, are of low commensurate international interest as a tourist destination. This may be a function of climate and geography—long-range flights are typically required for most non-US tourists.

Tourist numbers are not directly related to immigration levels; however, Canada's cities do not appear to engender as large a role in the global mindset as would seem to be warranted from their globalized nature and size. Many cities are pushing back on tourism as residents believe their quality of life is diminished by the crowds, e.g., Barcelona. This is not yet the case with Canadian cities.

Canadian cities also have much less direct benefit from tourism than their international comparators. The 2024 Taylor Swift Eras tour provides a powerful example (Sect. 3.4). The concerts in Toronto (6 nights) and Vancouver (3 nights) likely cost the municipalities of Toronto and Vancouver in overtime police and EMT staffing, and logistics support, exceeded the relatively modest contribution from increased hotel occupancy taxes. The provinces of Ontario and British Columbia, on the other hand, collected about $45 and $20 million respectively in sales tax, and the Federal Government about $85 million. Compare this to Chicago, where Taylor Swift played at Soldier Field June 3, 4, and 5 providing a net benefit to the City's budget of about $24 million (and $39 million to the State; $1 million to the federal government) [23].

5.4 Cities and the Jevons Paradox

In 1865, William Stanley Jevons, an English economist, observed that the rising efficiency of steam engines was, counterintuitively, increasing the amount of coal being consumed. As steam engines grew more efficient, people were finding new ways to use them. The Jevons paradox and the rebound effect, of increased consumption from lower prices, can exceed the savings from the efficiency gains.

In urban transportation, the Jevons paradox can be exacerbated through induced demand. When proposing a highway expansion, advocates often claim the project will reduce emissions along with congestion because car engines are more efficient (and less polluting) when traveling at optimum speeds. Because of induced demand however, the potential drivers who were forgoing the road because of congestion may be enticed back to the widened or new road. A Jevons paradox arises when the added

5.4 Cities and the Jevons Paradox

Table 5.4 Globalization and World Cities (GaWC) City Ranking [75]

Rank in 2010	City	Rank in 2020	City
1	New York	1	London
2	London	2	New York
3	Tokyo	3	Hong Kong
4	Paris	4	Singapore
5	Hong Kong	5	Shanghai
6	Madrid	6	Beijing
7	Beijing	7	Dubai
8	Seoul	8	Paris
9	Chicago	9	Tokyo
10	Singapore	10	Sydney
11	**Toronto**	11	Los Angeles
12	Sydney	12	**Toronto**
13	Shanghai	13	Mumbai
14	Milan	14	Amsterdam
15	Los Angeles	15	Milan
16	Moscow	16	Frankfurt
17	Washington	17	Mexico City
18	Brussels	18	Sao Paulo
19	Taipei	19	Chicago
20	Amsterdam	20	Kuala Lumpur

In 2010, Montreal (49) Vancouver (74), and Calgary (78) in addition to Toronto were the Canadian cities in top 100

In 2020, Montreal (41) and Vancouver (70) in addition to Toronto were the Canadian cities in top 100. Other Canadian cities were Calgary—105, Ottawa—174, Edmonton—219, Winnipeg—307, Halifax—312, Quebec—336, and Saskatoon—346

Created at Loughborough University, UK, the (GaWC) Research Network focuses on the external relations of world cities. Using four global service areas of accountancy, advertising, finance, and law, the index ranks connectivity of cities. A city-centered ranking on world flows between cities is provided, rather than the more typical state-centered world of boundaries.

Cities are classified into levels of world city network integration. London and New York consistently rank at the top as "alpha+" cities. These are complemented by "alpha+" that are mostly in the Asia/Pacific Region. "Alpha" cities are the remaining cities that are very important world cities that link major economic regions and states into the world economy. Toronto is consistently ranked as an alpha city.

Table 5.4 provides the top 20 globally connected cities in 2010 and 2020 as ranked by GaWC. In 2000, 15 of the world's top 20 connected cities were in OECD-member countries. This declined to 13 in 2010, and 11 in 2020. Toronto was ranked 10th in 2000, dropping one level each decade after.

emissions from those new cars, exceed the emissions saved from the vehicles that are no longer stuck in gridlock. In most Canadian cities, especially Toronto, traffic will not flow faster on the improved highway, due to induced demand. Electric vehicles will reduce emissions but do nothing about congestion.

Urban transportation, or better stated, urban mobility, requires a systems approach where modes of mobility and destinations can be disaggregated. Following a hierarchy is also sensible, with an effort to maximize walking (for health and mobility benefits) and cycling (a bicycle is several orders of magnitude more efficient in moving the rider than a train, bus or automobile). Transit (bus and rail) is the next best option, followed by ride sharing.

5.5 The Values of Gold: Resource Demands of Canadian Cities

That thin gold ring generated at least 20 tons of mine waste on its journey from the earth to finger. If there's a diamond on it another 40 tons of mining waste were created[3]. In addition to the mine tailings that often threaten waterways and leach toxins, that same ring generated about 350 kg CO_2e[4] of GHG emissions and required about 6 m^3 of process water[5] [24, 25].

In the rush to transition and decarbonize the world's energy systems, the importance of critical minerals is rising. Metals like cobalt, lithium, copper, rare earths like neodymium and cerium are increasingly needed. But what about gold? Canada's resource industry is inextricably linked to the metal. Canada is the world's fourth-largest gold producer.

Only about 10% of the world's gold is used for critical purposes like corrosion-resistant solder in electronics. The rest, 90%, goes to jewelry, coins, and bullion. Most of this is not critical to society, but rather reflects the value placed on an intangible material.

Diamond companies worry most about the changing preferences of potential customers. A diamond may be forever, but its value is transitory. In the 1930s, De Beers suggested spending one month's salary on an engagement ring. By the 1980s in the USA and Canada, De Beers increased their recommendation to two month's salary. And in some markets like Japan, De Beers even promoted spending three months' salary on an engagement ring[6] [26].

---

[3] From US Geological Survey, an engagement ring diamond is the product of the removal and processing of 200–400 million times its volume of rock; approximately 15 m^3 mine tailings, or 40 tons

[4] Canada's current (territorial) per person GHG emissions is about 20t per year and needs to be reduced to less than 2t per year by 2050, to credibly approach 'net-zero' targets.

[5] Estimates suggest that more than 95 percent of GHG, mine tailings and water use from diamond production can be reduced through lab-growing diamonds.

[6] See, The Atlantic How an Ad Campaign Invented the Diamond Engagement Ring (2015)

Prior to the 1930s, presenting a woman with a diamond engagement ring was rare (less than 10%). But by the end of the twentieth century, after successful marketing, 80% of future brides in OECD-member countries receive a diamond engagement ring [26]. The price per carat of fine diamonds rose from $2700 in 1960 to over $32,000 in 2022. Since then, prices dropped 25% from the peak, largely due to the Ukraine conflict and the growing impact of laboratory manufactured diamonds. Gemstone diamonds provide most of the industry's profits while 70% of diamonds are used as bort in industrial applications [27].

About 213,000 tons of gold have been mined throughout history, of which more than two-thirds has been mined since 1950 (all the world's mined gold would measure a cube of 22 m on each side) [28]. The price of gold stayed remarkably constant for 200 years from 1717 when Sir Isaac Newton, with the UK Mint, set the price around $18.95 per troy ounce [29]. Britain stopped using the gold standard in 1931, and the USA followed in 1933, with final abandonment of the standard in 1973. In the last 50 years, the price of gold has fluctuated from $36 in 1970 to $2650 in 2024. The price of gold rocketed from $36 an ounce in the early 1970s to $600 in the early 1980s, not necessarily reflecting a change in the value of gold, but rather to reflect the declining value of the US dollar as the country abandoned fixed-exchange rates linked to the gold standard [30].

Gold mining is shifting to ore bodies with declining grades, requiring more energy to mine and process, and leading to more environmental impact. Landfills typically now have higher grades of gold and other precious metals from discarded electronics than many new mines. New sources of metals being touted include asteroids and the ocean seabed. Future development of these resources, which carry a much larger potential environmental and social impact, will likely need more robust proof that the material is genuinely needed by society, and that the potential impacts are being aggressively minimized, *e.g.*, recycling through a circular process. Efforts are needed to ensure that potential lower and higher, prices do not equate to more overall environmental damage (see Sect. 5.4 for Cities and the Jevons Paradox).

5.6 Resources Canada

Canada is blessed by geography and challenged by history. Geography afforded the country abundant water, hydrocarbons, and minerals. Historically the country owes much of its wealth, especially within the common national narrative, to the extraction and sale of these abundant resources. Geography further blessed the country with a temperate climate and relative geopolitical security with three coasts and only one contiguous country that happens to be the world's most powerful country and a good neighbor.

Provinces were intent to generate wealth through resource development: electricity (mainly hydro), minerals, oil and gas, and agricultural exports. Much of this arose from the country's unique constitutional framework, allocating resource development to the provinces, and prescribing very limited revenue generation and legal standing to cities.

The main cities are 'sliced and diced' by a complex fabric of local governments, often at odds with their neighboring municipalities, regional governments, and service utilities. Mobility, a critical urban service, for example, is not coordinated in any of the country's five largest urban areas.

Not surprisingly, Canada's cities emerged as the most energy-intensive and wasteful in the world. Local governments, reflecting the national psyche, as well as maximizing limited fiscal tools, tend to encourage land development and new housing. Every local government in the five largest urban areas has several offices of economic development, charged with attracting businesses and developing land.

Shifting Canada to a less resource-intensive conserver society will be enormously difficult. Two system changes are required. First, the underpinnings of the country's economy are now based on resource development and a mindset of growth, i.e., selling more. No government—local, provincial, or national—can easily champion constraint and conservation. Recent federal and provincial support of electric vehicle manufacturing is a powerful illustration of this. A wholesale shift from internal combustion engines to EVs is supported with more than $57 billion in government subsidies [31], but the vehicles are still largely single occupant. From a mobility and quality of life perspective, these funds might better be allocated to integrated mobility applications; however, perceived economic priorities take precedence over local quality of life.

The second system change required to help move the country's cities to less resource intensity is the design of cities themselves. As much as 30% of new land development is devoted to automobiles, and already Canadian cities have some of the longest commute times in the world.

In 2022, every Canadian on average used more than 100,000 kWh of primary energy. An amount higher than almost any other country. For comparison, per person energy use in the USA was 78,754 kWh, China 31,051 kWh, UK 30,098, and India 7143 kWh. Canada's energy use is twice that of the average for people in other high-income countries. Almost four-times that of upper-middle income countries, and 83 times someone living in a low-income country. A Canadian uses more energy in a day-and-a-half than the nearly 100 million people in the Democratic Republic of Congo use in an entire year.

High energy use in Canada is reflected in the country's substantial greenhouse gas (GHG) emissions. Despite Canada's abundance of low-carbon hydro and nuclear electricity, annual per person GHG emissions are about 20 tons CO2e, about 4 times the global average, and more than 10-times those of people living in low-income countries.

Nature abhors a vacuum; it also makes every effort to move higher concentrations to areas with lower concentration. The world's weather patterns and ocean currents, for example, are mostly about moving excess heat from the equator to cooler climes north and south. Heat, like water, eventually, always flows downhill.

5.6 Resources Canada

Canada's energy and resource consumption, being as high as they are, relative to the rest of the world, suggest similar averaging pressures should be anticipated. The Canadian Confederation is predicated largely on equalization efforts by the federal government. Efforts are made through fiscal transfers to assure that levels of services and quality of life are roughly similar in all parts of the country. Significant wealth is transferred from urban areas to rural areas, and from more affluent provinces like Alberta.

Equalizing wealth in a country as disparate as Canada is always a work in progress. Obfuscation, resentment, borders, and barriers arise. Why should a rural resident's lifestyle be subsidized? Why should someone in the city have anything to say about the lifestyle of rural residents? Why should Alberta subsidize Quebec, especially when Quebec seems unhappy with even being in Canada?

Canada of the late twentieth century will likely represent the apogee of energy- and materials-intensive societies. As the world decarbonizes and shifts toward a more circular (conserver) economy, Canadians should brace for similar equalization pressures between higher-income Canadians and the world's poor. Borders can serve to delay these pressures, but they can rarely do so forever.

Canada's high energy and material use is driven by three national attributes: urban form with its high reliance on the automobile and large, distant homes; an energy-intensive industrial base, such as extraction of bitumen from the oil sands; and a mindset of abundance typified by the world's least fuel-efficient vehicle fleet, and communities with no water meters and water losses in excess of 30%.

Energy is only five% of GDP, but it is the first five percent. In a hierarchy of needs for cities, lifestyles, education, health care, and the urban economy are all atop a foundation of energy, food, and water.

"Canada is a big, cold country" is often a reason given for high energy demand and very high per capita GHG emissions; however, Canada's GHG emissions should be relatively easy to reduce. Eighty percent of the country's population lives in reasonably well-managed cities with ample space for geothermal and deep lake/ocean cooling and heating. Toronto's world-class Enwave, which supplies heating and cooling to more than 180 buildings, could readily be replicated in Montreal, Vancouver, Halifax, and other cities.

About 90% of Canada's population lives in a thin band less than 200 km from the US border. Connecting these communities by rail and frequent bus transit would again, be a relatively easy engineered solution. For roughly the same cost of existing highway maintenance, high-speed, frequent 'surface subways' could be added to the country's main highways. High-speed rail may be more appealing than buses; however, this is much more expensive, and in the meantime, with dedicated lanes, frequent service, and integrated ride sharing and ride hailing, more than half the urban commuting (inter and intra) could be shifted to low-carbon alternatives. A preliminary mobility as a service review of southern Ontario suggested a possible annual savings of $8000 per household [32]. Despite the potential environmental and economic benefits, this is a challenge to implement, as it requires broad and sustained collaboration, shifting away from the primacy of auto-manufacturing, and sustained community trust that the service will be safe, reliable, convenient, and affordable.

Canadians by virtue of their resource-based economy, typically support mining and fossil fuel extraction. Efforts to maximize supply underpin much of the country's history and wealth. Support varies by material and region. Quebec, with few reserves is less supportive of oil and gas, but far more supportive of large-scale hydroelectric development. Governments provided infrastructure to facilitate resource extraction: highways, railways, seaways, pipelines, electricity generation, and distribution.

As early as 1670 with the founding of the Hudson Bay Company and Europe's demand for beaver pelts, there has been a conflict within the psyche of Canadians. Fossil fuels, trees, minerals, and beavers; Canada's resource sector maximized supply, despite growing awareness of impacts, eventual limits, and externalities. Before European settlers arrived, Canada's beaver population was estimated at 60 million. The population plummeted to as few as 100,000 animals, and only avoided extirpation by changing fashions and early 20th-century conservation efforts [33].

Asbestos was declared a human carcinogen by the World Health Organization in 1987 [34]. However long after the health problems of asbestos were proven, the Government of Canada continued to aggressively work with the Province of Quebec to market the material internationally (while having its use banned in Canada, establishing a marketing group in Montreal, and changing the name to chrysotile to help marketing efforts) [35]. In 2004, Canada blocked the Rotterdam Convention attempting to limit the use of chrysotile asbestos [35], and it was not until 2018 that the Government of Canada finally took efforts to fully ban the use of asbestos (Iceland first banned asbestos in 1983, and more than 108 countries restricted asbestos before Canada acted) [36, 37].

Canadians may be quick to support in principle the shift to a circular economy; however, meaningful action often lags the rest of the world. More power plants, more timber and building materials, more oil and gas can always be found, and if more recycling, and greater circularity, is needed to make that possible, so be it. In the next 50 years though, Canadian cities will need to transition with the rest of the world to more efficient, more resilient, less energy, and materials-intensive cities. Canada's cities will need to lead the shift in focus from the emphasis over the last 100 years on increasing resource supply to lowering demand. Constitutionally, the provinces will not lead these efforts, nor the federal government as it continues to balance conflicting regional demands.

Canada's infrastructure history is underpinned by governments providing infrastructure to facilitate resource extraction: highways, railways, seaways, pipelines, electricity generation, and distribution.

The first 100 years of Canada's urban phase were mostly about defining what kind of country would emerge. Newfoundland and Labrador would be included, Quebec too, the vast north would be governed to reflect more closely the rights of Inuit and impacts of an opening Arctic. The influence of provinces and regions will always ebb and flow. Ideas and cultures, like social tectonics, will continue to shift and abrade.

Influencing resource demand was less top-of-mind in Canada. This is especially true in the design of cities. Canada's cities, intentional or not, were developed to use much of what Canada was offering, cheap electricity, oil and gas, automobile manufacturing, land development and an active building industry.

5.6 Resources Canada

No country would find 173 billion barrels of oil in the ground and leave them there.
—Prime Minister Justin Trudeau

Previously, no country would find billions of barrels of oil in the ground and leave them there, but at some point, they will. Availability of supply will not determine how much oil is consumed. Rather, demand will determine final amounts of oil, and gold, and most materials consumed, or left in the ground.

Peak oil is that future date when the world's use of oil will reach a peak, after which time use will decline. The date of the peak, its amplitude, and the rate of decline post-peak are the critical questions. Answers to these questions will come from customers, most who live in cities. The date of peak oil is now closely observed and hotly debated. The International Energy Agency predicts peak oil will be reached before 2030 [38]. The marketing pressures of regions like Alberta, companies like Exxon claiming the peak is beyond 2050 [39], and the second Trump Administration's desire to take the USA out of the Paris Agreement, may delay the peak; however, as the price of renewables and storage continues to decline, and the externalities of human health and climate change are better internalized in the price of fossil fuels, the demand for oil (with gas and coal) will steadily decline.

A city perspective of supply and demand for resources is critical. Beavers were largely extirpated in Canada; however, it was the change in European (urban) fashions that saved them from extinction. So too most species of whales. Whale oil for lighting and baleen for fashion, almost wiped out the world's whales, but demand plummeted as kerosene replaced whale oil, and they received a reprieve. Bison were not so lucky—they were virtually exterminated due to possibly purposeful over-hunting and habitat destruction [40–42].

Like how cities need to take a more fulsome accounting of GHG emissions, a global accounting for the impacts of materials used in the city as it relates to biodiversity loss, land-use change, and other pollutants, is necessary.

Cities and megaregions will need to host challenging conversations on degrowth, value, equity, sustainable development, reconciliation, planetary boundaries (limits), net-zero carbon emissions, and myriad shifts in the next decades. These will be difficult and may be at odds with senior levels of government, local businesses, and residents. Cities may elect to delegate much of the conversation to city associations such as FCM; however as Chap. 6 outlines, Canadian cities like Toronto Region, Ottawa-Gatineau, Montreal and Vancouver CMAs, and Calgary-Edmonton need place-based sustainability planning. Most of the objectives will be common across Canadian cities; however, locally specific approaches that account for context are required, along with community established metrics and targets.

Indications of how difficult these discussions might be, are illustrated through recent public delegations to municipal councils petitioning against 15-min cities, e.g., Edmonton, Milton and Pickering [43], and Cleveland in the USA [44]. Municipal officials are surprised at the vehement opposition to the concept (more accessible communities), and in what their minds is the absence of facts and perspective in the opposition. A similar concern was raised by municipalities in the hostility expressed against vaccines.

Canada's path to net-zero carbon, as stipulated through the Canadian Net-Zero Emissions Accountability Act, which came into force on June29, 2021, to move Canada's economy to net-zero greenhouse gas emissions by 2050 [45], will be more difficult than most countries for at least three reasons. First, Canada needs to reduce energy demand and emissions from a much higher per capita level than most countries. Second, targets like those under the IPCC Paris Agreement are total emissions by country. Canada with higher rates of immigration is the OECD's fastest-growing country. Relative per-person emissions will need to fall faster than any other country.

Lastly, Canada also has very different climate mitigation strategies by region. Alberta, for example, will be challenged with the global energy transition if fossil fuel markets decline as quickly as suggested by the IEA. Quebec, on the other hand, with its supply of low-carbon electricity will pursue electric vehicles and long-term (low carbon) power agreements with neighboring jurisdictions.

The complexity of Canada's climate mitigation efforts is exacerbated when provinces and regions propose alternative pathways. These pathways will be multi-layered as climate change mitigation is only one of many forces acting on Canadian regions. The seventh planetary boundary (ocean acidification) is soon-to-be crossed (adding to the already crossed boundaries of climate change, biodiversity losses, nitrogen and phosphorus use, freshwater use, novel entities, and land-use change).

Promoting resources by province or region, e.g., hydrocarbons from Alberta or hydroelectricity from Quebec, will vary. Cooperation is needed. For example, landlocked Alberta's need to ship product across provincial borders, or excess hydroelectricity from Newfoundland and Labrador. Quebec may have relatively low-carbon emissions; however, aggrieved provinces will be quick to point out profligate water use in communities like Montreal (Montreal is one of the few cities that does not meter residential water use) [46].

Shifts in resource development plans may drive urban and rural Ontario farther apart. Or as the need for climate adaptation grows faster in northern Canada, will the country's majority in the south be willing to pay for infrastructure enhancements?

New approaches are needed to ensure consideration of an aggregated voice of regional communities like Yellowknife, Winnipeg, and Nunavut, as well as the perspective of larger urban centers (Montreal, Ottawa-Gatineau, Toronto, Calgary-Edmonton, Vancouver), even when, or especially when, those voices differ from their host province, or the federal government.

5.7 Leave It on or in the Ground: Beyond Beavers, Bison, and Oil

"Could we vault gold in the ground, creating a completely different investment class?" Michelle Ash asked in a 2017 IdeaCity address. "We [could] start using blockchain technologies, so that we don't have to mine at all," she said, noting that two-thirds of the gold produced is used as a "financial instrument."[7]

Ms. Ash who was Chief Innovation Officer of Barrick Gold Corporation in Toronto at the time, claimed that Barrick seriously explored doing this—"creating social and economic value from gold that's still in the ground."

But the project was abandoned in the aftermath of Barrick's 2019 merger with Randgold (explained Ms. Ash, now Vice President of Growth at BHP). At least one junior company is trying to move this model forward, with a gold deposit in Northern Ontario, but it remains early days.

For Barrick, this was part of a wider experiment to see if Canadians of Indian heritage would buy, sell, and trade electronic credits representing real units of gold—e-gold—with their relatives in India. Indians buy more gold jewelry than anyone else on Earth–China is second. The gold is not just jewelry but represents a store of wealth that women have traditionally brought to a marriage partnership. People were comfortable with buying and selling e-gold as a proxy for the real thing.

Mining for diamonds could also be curtailed. Diamond mining typically brings with it environmental damage and human rights and worker safety issues. Diamonds can now be created in a laboratory by placing carbon in a pressure chamber (although CO_2 emissions may be higher).

"The worth of gold is not in its utility" [47]. The same is true for gemstones like diamonds and emeralds. The social license and intrinsic value for these materials are likely to decline as utility does not warrant the environmental impact (i.e., the externalities will be better internalized).

This *irrational exuberance* of placing value in gold, gemstones, or crypto-currency was exhibited as early as the Dutch Golden Age in the seventeenth century when a single tulip bulb could fetch a price as high as an upper-end house in Amsterdam [48]. Prices reached a frenzy between 1633, when the market was opened to non-professionals, and 1637, when prices plummeted, leaving many destitute [49].

Over the next 50 years as the environmental and social costs of materials are better defined, the non-utilitarian fraction of the item will be more challenging to defend if impacts on the commons are high. Greater efficiency of use will also be expected. There is nuance in the definitions and measurement, and industry associations will promote their products and protest any form of limits. The shift will be led by residents of cities as they are the main customers. Discouraging 'blood diamonds' is an early manifestation of this changing social license for materials. The reprieve Canada's beavers received when fashion changed in London is an earlier example.

---

[7] Pollon, P. (2023). This discussion on leaving gold in the ground is summarized from *The case for leaving gold in the ground* (Globe and Mail, Oct 6, 2023) and Pitfall, The Race to Mine the World's Most Vulnerable Places (2023).

On August 26, 2021, downtown Wheatley, Ontario, was wracked by a massive explosion. Twenty people were injured, and several buildings were damaged and destroyed [50]. The community's economy was damaged as much of the downtown core was closed for more than four years.

The Wheatley explosion was caused by leaking gas from a long-abandoned wooden-cased petroleum well [51]. There are 27,000 known petroleum wells in Southern Ontario [52], where the world's petroleum industry got much of its start in 1858 near Oil Springs, Ontario.

Those 27,000 petroleum wells, plus perhaps that many again that are unknown in Ontario, as well as another 250,000 abandoned and orphaned wells in Alberta and Saskatchewan, are a powerful and problematic legacy. In addition to the dangers of possible explosions and other human health impacts, many of the wells are leaking methane and hydrogen sulfide gas into the atmosphere.

In Alberta alone, the unfunded liability to remediate these abandoned wells is likely more than \$260 billion [53, 54]. Previous estimates of the fugitive emissions from these wells and their contribution to climate change (CH_4 and H_2S) are significantly under-estimated [55]. Approximately 43% of Canada's anthropogenic methane emissions originate from oil and gas systems, and about 42% of that is from fugitive emissions (at least 75 Mt in 2018) [56].

Abandoned oil and gas wells in Canada are only one of the many resource liabilities left to communities. Natural Resources Canada estimates that there are more than 10,000 orphaned and abandoned mining sites across Canada [57]. The legacy liabilities of several abandoned mines are enormous. For example, the Giant Gold Mine in Yellowknife with more than 237,000 t of arsenic on site is expected to cost the Government of Canada more than \$4.4 billion to remediate. The federal government was forced to take responsibility for environmental liabilities in 1999, when Royal Oak Mines Inc. went into receivership. Remediation began in 2022 and is expected to continue until 2038, after which the site will still require perpetual care and maintenance.

In addition to the 237,000 t of arsenic located on the abandoned Giant Gold Mine site in Yellowknife, several decades of metal processing activities caused widespread accumulation of arsenic in biomass, soils, and sediments in neighboring landscapes. The 2023 forest fires that burned 2500 km^2 around Yellowknife alone released between 15 and 59% of global annual arsenic wildfire emissions, representing between 2 and 9% of total global arsenic emissions (between 69 and 183 tons) [58].

Leaving valuable resources in the ground (or on the seabed) is a disruptive idea. The mining industry will likely not lead the charge to reduce mining. Nor will oil and gas companies push for an earlier and lower peak oil. However, as Barrick initially highlighted, the transition could happen much faster with the active support of energy and resource companies.

As the costs to remediate abandoned resource development sites across Canada are more clearly known (perhaps as high as \$350 billion for oil and gas, and another

$250 billion for mining activities)[8], and as health impacts are better understood, as well as better inclusion of the externalities of resource development such as climate change and biodiversity loss, the social license will likely grow more tenuous. Much of this debate will occur in Canadian cities. If conservation is to flourish, it will need to be anchored in the buying-habits of urban residents, including the stocks and bonds they invest in, and the resource development rights they entrust to their governments.

The world needs metals, but the world does not always need more mining. Progressive mining and energy companies will shift from outputs, such as tons of metal produced, to outcomes, such as societal utility provided.

As outlined in Sect. 6.3 through discussions with Mark Cutifani, Chairman of Vale Base Metals, more progressive resource companies are likely to develop new financing structures that pay for the benefits (and presumably more of the externalities) of energy and materials, rather than the commodity itself. Active support of this transition by the resource company customers, and shareholders (and host cities), will speed the transition.

5.8 Blame Canada, Rich Canadians, or All the Rich?

Section 5.1 outlines the findings from UNEP's International Resource Panel that provide several decades of data on Canada's energy and material production and consumption. Canada produces, uses domestically and exports, significant quantities of energy (petroleum, coal, hydroelectricity, uranium), and materials (potash, metals, beef, grains, and manufactured products). A few countries like Australia and Saudi Arabia might produce more resources per person, but Canada is among the world's largest. Canadians, largely as a function of the spatial development of their cities, also use more energy and materials than almost any other country (per person). A few countries like Norway might use more electricity per person, or Bahrain that uses more petroleum, but no major country uses more primary energy per person than Canada. Canadians are also relatively affluent, so all residents, urban and rural (especially those in isolated communities) are some of the world's largest consumers of energy and material resources.

Global convention is that most energy and material production and consumption data is aggregated and published by country. The UN, OECD, and World Bank, institutions representative to member countries for example, provide the best energy and material (and GHG emissions) data. This makes sense as many international treaties, such as climate change within the UNFCCC, are negotiated by country; however, nuances are lost. For example, total residential GHG emissions of two neighborhoods

---

[8] Author estimate (based on a likely minimum cost for Wheatly, ON remediation of $35 million; more than 300,000 abandoned oil and gas wells in Canada; Alberta Auditor General's rough estimate of $260 billion for that province's well remediation liabilities; several mine site remediation projects already more than $1 billion in remediation costs per site, e.g. Giant Gold Mine, Faro Mine, Port Hope Initiative; more than 10,000 abandoned mining sites)

in the same community (Toronto Region) can vary from 1.31 tons/person (East York) to 13.02 tons/person (Whitby) [59]. Or that Alberta's GHG emissions are 59.8 CO_2e tons/person [60] while Quebec's are about 10 tons CO_2e [61] (Canadian average 18.2 tons CO_2e).

A Canadian family of four that takes a cruise every year, has two vehicles, a large dog, big house in the suburbs, cottage, and eat lots of meat, likely have global GHG emissions over 150 tons CO_2e per year. A similarly affluent Canadian family living in downtown Vancouver or Montreal, with a conserver lifestyle, could likely keep their GHG emissions below 50 tons per year. One family is not better than the other, however in Canada's overall GHG emissions, both are averaged into the national total. Caution is also needed in that GHG emissions are only one issue of concern from an earth system (e.g., biodiversity, land use) or socio-economic perspective (economy, equity).

Tian et al.(2024) provide a different perspective in *Keeping the global consumption within the planetary boundaries* [62] where they ascertain specific responsibility on six environmental indicators by the world's income groups (GHG emissions, biosphere, freshwater use, phosphorous and nitrogen flows, and land use). Using expenditure databases, they estimated total global environmental impacts by global consumer decile. Tian et al. divide the world into the ten wealthiest to poorest groups, irrespective of where they live.[9] They show that 31–67% and 51–91% of responsibility for global environmental impacts could be attributed to the global top 10 and 20% of consumers respectively. The world's wealthiest 20% of population are responsible for more than 63% of all CO_2 emissions (43% attributed to the wealthiest 10%; 14% to the wealthiest 1% of people).

Modifying the lifestyles of the world's wealthiest ten percent would yield the fastest most comprehensive benefit to earth systems. Seventy percent of Canadians have sufficient wealth to place them the world's wealthiest decile [63].

The top 1% of global population holds about 50% of the world's total wealth (with a net worth of more than US$1.3 million). The top 10% holds approximately 85% of global wealth [64]. The 2024 UBS Global Wealth Report estimates that 39.5% of world population has less than $10,000 in assets. 42.7% of population has wealth between $10,000 and $100,000; 16.3% between $100,000 and $1 million; and 1.5% over $1 million [65]. In 2022, there were an estimated 59.4 million millionaires, which is expected to rise to 86 million in 2027. In 2023, Canada was ranked 10th in average wealth (US$375,000 per adult; $142,587 median). Globally, in 2023 there were 2,664 individuals with wealth more than $1 billion (67 in Canada [66]). Nothing would reduce GHG emissions faster than changing the lifestyles of these 2664 people.

---

[9] Affluence can be measured by either wealth (owned assets) or income. Wealth is a better measure as the world's richest may have a relatively lower income, compared to their wealth. Tian et al. disaggregate affluence by levels of consumption, which provides a comprehensive measure of environmental impact.

Box 5.1 Scope of the Challenge, a Consumer's Tale

Doug is about to propose to his soon-to-be (hopefully) fiancé. In preparing for the big night, he did his homework.

His girlfriend, Dana, is fiercely determined to reduce greenhouse gas (GHG) emissions. Of course, that's one of the things he loves about her. Doug's cooked a special meal. Chicken, quinoa, and spinach salad, with apple pie for dessert. They have not given up meat but have reduced the amount they eat and try to stick to poultry and some fish (the whole meal, he estimates, is about 5 kg CO_2, plus 1.5 kg CO_2 for the bottle of good French wine). Doug also bought a dozen roses (about 2 kg CO_2 per stem[10]).

The engagement ring is another 550 kg CO_2 (350 kg for the gold, 200 kg for the 1 carat diamond).

The total annual GHG emissions of a Canadian are around 20 t (20,000 kg CO_2 per year) and Canada has committed to get to zero by 2050. The engagement ring and their wedding (probably another 1000 kg CO_2) are a special occasion that the couple hope will be offset over their lifetime choices, especially when alternatives, like better transit, EVs, and low-carbon energy (carbon equivalent of less than 20 g/kWh), are the norm, which will make reducing emissions easier.

Doug can calculate the GHG emissions associated with the big night because of efforts made by a few businesses and the World Business Council for Sustainable Development (WBCSD) and World Resources Institute (WRI). Work started in 1997, and in 2001, a global GHG emissions protocol was launched.

For business to track their emissions, they needed a more comprehensive method than the territorial approach countries use to determine national emissions (estimate all emissions generated within the country's borders). The same is true for Doug wondering about his ring and roses. The gold may have come from Canada, the diamond from Botswana, and roses from Ecuador, but he wants to know the total emissions, especially the upstream (Scope 3) emissions.

The WBCSD-WRI emissions inventory uses Scopes 1, 2, and 3 to account for where emissions are generated, while reducing the risk of double-counting. Scopes 1 and 2 are direct emissions, such as electricity and fuel used in company vehicles. Scope 3 is embodied emissions. For example, the upstream emissions in the parts a vehicle manufacturer assembles, or the downstream emissions from the eventual use of that manufactured vehicle after it is sold.[11]

The larger challenge that Doug and Dana face is that GHG emissions are only one issue. The rate of biodiversity loss also needs to slow dramatically, so

---

[10] In Canada, most roses are imported from South America (typically Columbia and Ecuador) and driven north from the main US Airport, Miami or Los Angeles. Emissions estimated to be slightly lower than those for roses in the UK (Dutch Roses: 2.437 kg CO_2, Kenyan roses: 2.407 kg CO_2–see www.flowersfromtheuk.org).

[11] See ISO 14064-I: Greenhouse Gas Protocol.

too air and water pollution. A healthy economy in Canada and globally is also important, with safe working conditions for everyone. The couple have talked about having one, maybe two, children. What kind of planet will they inherit, and what's the impact of bringing two more Canadians into the world?

The engagement ring generated 60 tons of mining waste, and what if it is a 'blood diamond' (despite the Jeweler's assurances it was not). The complexity of the challenge—accounting for the impacts at all stages of human activities—is significant.

The halting progress in GHG accounting is a good example of the challenge. In recent years, more businesses started to provide comprehensive Scopes 1, 2, and 3 (upstream and downstream) emissions, but many stopped and delayed, claiming customers were not interested or the process was too complicated, especially for Scope 3 emissions. Municipalities, wanting to track their GHG emissions and progress toward net-zero targets, also started to track GHG emissions. However, only a few include Scope 3 emissions. As Doug's roses and ring highlight, without accounting for upstream and downstream impacts, the inventory is likely woefully inadequate.

Many businesses and governments expanded GHG emissions accounting efforts to include broader environmental, social, and governance (ESG) issues. Alberta's claims that their oil and gas are preferable to say, Saudi Arabia's, with a poorer human rights record is illustrative. In response, Saudi Arabia may question Alberta's history with First Nations communities or ask what the equivalency is between higher GHG emissions versus human rights. These complexities are behind many of the recent delays in corporate reporting of ESG metrics and postponing of Scope 3 accounting. There may also be a way to put off meaningful mitigation.

Dana and Doug's household, and the community they live in, is the optimum scale for GHG inventories and ESG accounting. A business, for example, is most concerned with selling their product. ESG metrics might be a way to differentiate themselves from a competitor, but they will be reluctant to provide ongoing information if it suggests reducing sales. Most national and provincial governments are the same. Economies and GDP are largely derived from the sale of resources and manufactured products.

Households and communities on the other hand are concerned with both maintaining the economy (for jobs and GDP for example), as well as reducing local and global consumption (for reduced planetary impacts). The balance between the two and the pace of environmental remediation and social equity are key.

The Scopes 1, 2, and 3 approach, with an appreciation for upstream and downstream impacts, can be used for activities beyond GHG emissions. A good faith account of value chain impacts will help people like Dana and Doug to make the right choices for their families and the health of the planet.

5.8 Blame Canada, Rich Canadians, or All the Rich?

Box 5.2 A Measure of Success; US Embassies and Local Air Quality Around the World

In 2008, the US Embassy in Beijing started monitoring PM2.5 at the embassy. Local US citizens were given access to the ambient air pollution information and acted accordingly when they could. The program was considered a success, and by 2019, the US State Department was providing local air-quality data in 43 cities in 27 countries [67]. The data was tweeted in real time, making it easily available to US citizens, embassy staff, and the broader public. Residents also saw the information and changed their behavior, or urged community representatives to act.

In 2012, one out of every nine deaths worldwide was attributable to air pollution [68]. A 2021 report by Health Canada estimated that air pollution contributes annually to 13,500 premature deaths in Canada (costing $120 Bn in 2016) [69]. Could something as simple as providing information have a measurable impact on a challenge so large? Apparently so.

Starting in 2008 when air-quality data at the US Embassy in Beijing was tweeted hourly, international attention was drawn to poor air quality. By 2020, over 50 US diplomatic facilities in 38 countries were live-tweeting air-quality readings [70].

By raising public awareness and measuring the effectiveness of actual improvements, Jha and La Nauze (2022) found that this natural experiment saw a dramatic improvement in local air quality. The US embassy air-quality tweets led to global health benefits with annual monetized health benefits of $127 million per city. With relatively modest cost to implement, the public education program provided more than $6.3 billion in annual benefits, plus significant enhancement of local quality of life.

Box 5.3 The Second Waste Hierarchy. Politics Matters Most

People familiar with waste management know well the waste management hierarchy; reduce, reuse, recycle, recover. There is however a second, less well-known hierarchy on the politics of waste management. This is often distilled to 'where there's muck there's money', but money is only part of it. Fear and greed, and the more subtle forms of interference, obfuscation, delay, and excuses also mix with waste, and underpin the politics of communities.

Mayor LaGuardia, talking about New York City, is credited with saying that there is no Republican or Democratic way of taking out the garbage. True, partisanship can be reduced, but the politics of waste management is almost impossible to separate. Small 'p' politics is why Canada sends much of its waste to the USA for disposal, or why some communities spend $160 per ton to dispose of waste at an energy from waste (EFW) facility instead of landfilling for about $60 per ton. Politics is why Southern Ontario is fast approaching a disposal crunch with no new landfills being developed, and almost no one is talking about the impending crises.

Politics is why Canada's recycling rate has stubbornly stayed the same since the 1980s, and why beverage containers that make up less than 2% of the waste stream, drove much of the country's waste management planning, especially in Ontario. Politics is why most Canadian cities have limited knowledge of how much waste is generated within their communities, or where it is being disposed. Politics is the main reason Canadians generate more waste than just about anyone else in the world, and why relatively so little is recycled. Politics is why some rural communities have blue box recycling when the emissions and costs of collection vehicles are several orders higher than any potential benefits.

Organized crime often grows from the waste management industry [71]. For example, Chicago provides a rich history of the inner workings of groups like the mafia and the underworld of waste [72]. So too Italy's experience with hazardous waste [73]. Waste management is often shrouded in privacy and is a relatively small part of ongoing budgets. The elasticity of demand for waste removal is low. People will quickly pay to be rid of it, an optimum environment for organized crime and political pressures to take root.

Communities are tasked with at least two things, roads and garbage. A small-town mayor knows well that the roads should be resurfaced before an election, and that she will always receive complaints about snow removal and waste collection. However, a provincial premier and political party will eventually interfere in things like a city's bike lanes, as well as much of the mechanics of waste management. The political influence can be subtle; perhaps an industry lobby group extolling the virtues of a circular economy, with new jobs and tax revenues. Or the influence might be heated and visceral—a neighboring community group doing almost anything to fight a proposed landfill.

As organized crime and parochial politics grows out from streets and waste, civility and sustainability are also most likely to grow from the ground up. Equity, efficiency, transparency, good government and management, underpin best-practice waste and traffic management everywhere. Capacities that can be recycled throughout all urban services.

References

1. Byrnes H, Frohlich TC (2019) Wall Street 24/7. https://www.usatoday.com/story/money/2019/07/12/canada-united-states-worlds-biggest-producers-of-waste/39534923/
2. Hoornweg D, Bhada-Tata P (2012) What a waste: a global review of solid waste management. World Bank

References

3. Kaza S, Yao L, Bhada-Tata P, Van Woerden F (2018) What a waste 2.0: a global snapshot of solid waste management to 2050. World Bank Publications
4. https://ised-isde.canada.ca/site/canadian-automotive-industry/en. Accessed 27 Oct 2024
5. https://www.canadaaction.ca/potash-mining-canada-facts. Accessed 27 Oct 2024
6. United Nations Environment Programme (2024) Emissions Gap Report 2024: no more hot air … please! With a massive gap between rhetoric and reality, countries draft new climate commitments. Nairobi
7. www.unilever.ca
8. World Business Council for Sustainable Development (2011) WBCSD/WRI greenhouse gas protocol product standard adopted by the sustainability consortium, Geneva
9. Fraser A (2017). Are investment carbon footprints good for investors and the climate? Policy Options
10. https://www.conferenceboard.ca/hcp/municipal-waste-generation-aspx/. Accessed 25 Jul 2024
11. Mensah D, Karimi N, Ng KTW, Mahmud TS, Tang Y, Igoniko S (2023) Ranking Canadian waste management system efficiencies using three waste performance indicators. Environ Sci Pollut Res Int 30(17)
12. International Energy Agency (2018) Average personal vehicle characteristics, by country
13. Natural Resources Canada (2022) Minerals and the economy
14. United Nations Environment Programme (2024) Global Resources Outlook 2024: bend the trend—pathways to a livable planet as resource use spikes. International Resource Panel, Nairobi
15. https://w3.unece.org/SDG/en/Indicator?id=54. Accessed 22 Nov 2024
16. IRP (2018) The weight of cities: resource requirements of future urbanization. In: Swilling M et al (eds) A report by the international resource panel. UNEP, Nairobi, Kenya
17. Cabernard L, Pfister S, Hellweg S, Baptista MJ (2019) Natural resource use in the group of G20: status, trends, and solutions. UNEP International Resource Panel
18. Environment and Climate Change Canada (2020) National waste characterization report: the composition of Canadian residual municipal solid waste
19. Weichenthal S, Van Rijswijk D, Kulka R, You H, Van Ryswyk K, Willey J, Jessiman B et al (2015) The impact of a landfill fire on ambient air quality in the north: a case study in Iqaluit, Canada. Environ Res 142:46–50
20. IRP (2019) Global resources outlook 2019: Natural resources for the future we want; Oberle B, Bringezu S, Hatfeld-Dodds S, Hellweg S. A report of the international resource panel. United Nations Environment Programme, Nairobi, Kenya
21. WorldData.info. Accessed 1 Aug 2024
22. EuroMonitor (2023)
23. Hoornweg (2025) A sustainability assessment of the Taylor Swift Eras Tour (in preparation)
24. Sun Y, Jiang S, Wang S (2024) The environmental impacts and sustainable pathways of the global diamond industry. Human Soc Sci Commun 11(1):1–12
25. World Gold Council (2019) Gold and climate change: current and future impacts (diamond about 2/3 of the GHG emissions and water use; gold 1/3)
26. https://www.bbc.com/news/magazine-27371208. Accessed 28 June 2024
27. World Diamond Council (2024)
28. https://www.gold.org/goldhub/data/how-much-gold. Accessed 28 June 2024
29. World Gold Council (2024)
30. Conway E (2014) The summit: Bretton woods, 1944. Pegasus Books
31. https://thehub.ca/2024/04/25/40-billion-to-ev-automakers-good-investment-or-risky-gamble/?t&utm. Accessed 23 Nov 2024
32. Ontario Tech University (2018) Improved transportation in the Toronto region
33. Backhouse F (2013) Rethinking the beaver. Canadian Geographic
34. https://www.canada.ca/en/environment-climate-change/news/2018/10/the-government-of-canada-takes-measures-to-ban-asbestos-and-asbestoscontaining-products.html. Accessed 23 Nov 2024

35. Ruff K (2017) How Canada changed from exporting asbestos to banning asbestos: the challenges that had to be overcome. Int J Environ Res Public Health 14(10):1135
36. Lin RT, Chien LC, Jimba M, Furuya S, Takahashi K (2019) Implementation of national policies for a total asbestos ban: a global comparison. Lancet Planet Health 3(8):e341–e348
37. https://www.canada.ca/en/environment-climate-change/news/2018/10/the-government-of-canada-takes-measures-to-ban-asbestos-and-asbestoscontaining-products.html. 23 Nov 2024
38. IEA (2024). Oil 2024. Analysis and forecast to 2030
39. https://www.netzeroinvestor.net/. Accessed 23 Nov 2024
40. Feir DL, Gillezeau R, Jones ME (2024) The slaughter of the bison and reversal of fortunes on the Great Plains. Rev Econ Stud 91(3):1634–1670
41. Daschuk JW (2013) Clearing the plains: disease, politics of starvation, and the loss of Aboriginal life. University of Regina Press, vol 65
42. Holt SDS (2018) Reinterpreting the 1882 bison population collapse. Rangelands 40(4):106–114
43. Rider D (2024) The creator of the '15-minute city' talks about how he would combat people's fears of his idea, 27 Nov 2024. The Toronto Star
44. Carey C (2024) Why the 15-minute city continues to inspire municipal leaders. Cities Today
45. https://www.oag-bvg.gc.ca/internet/English/att__e_44374.html
46. Montreal mayor says no to water meters as city looks to reduce consumption. CBC News
47. Demuth B (2019) Floating coast: an environmental history of the Bering Strait. WW Norton & Company
48. Pavord A (2019) The tulip: the story of a flower that has made men mad. Bloomsbury Publishing
49. Dash M (2011) Tulipomania: the story of the world's most coveted flower and the extraordinary passions it aroused. Three Rivers Press
50. Engineering & Environmental Consulting Ltd. (2024) Wheatley, Ontario explosion. https://360eec.com/wp-content/uploads/2024/10/Case-Study-Wheatley-Ontario-Explosion.pdf
51. El Hachem K, Kang M (2022) Methane and hydrogen sulfide emissions from abandoned, active, and marginally producing oil and gas wells in Ontario,. Canada Sci Total Environ 823:153491
52. Ontario GeoHub (n.d.) https://geohub.lio.gov.on.ca/datasets/petroleum-well/explore?filters=eyJXRUxMX01PREUiOlsiUGx1Z2dlZCBiYWNrIGFuZCB3aGlwc3RvY2tlZCJdfQ%3D%3D&location=42.974254%2C-81.946816%2C8.00. Oil and Gas well map
53. Cosbey A (2022) The bottom line: why Canada is unlikely to sell the last barrel of oil. International Institute for Sustainable Development (IISD) (Report)
54. https://www.oag.ab.ca/wp-content/uploads/2023/03/Liability-management-oil-gas-mar2023.pdf. Accessed 24 Nov 2024
55. MacKay K, Lavoie M, Bourlon E et al (2021) Methane emissions from upstream oil and gas production in Canada are underestimated. Sci Rep 11:8041. https://doi.org/10.1038/s41598-021-87610-3
56. https://www.canada.ca/en/environment-climate-change/services/climate-change/greenhouse-gas-emissions/sources-sinks-executive-summary-2024.html
57. https://publications.gc.ca/collections/collection_2010/nrcan/M39-124-eng.pdf
58. Sutton OF, McCarter CPR, Waddington JM (2024) Globally-significant arsenic release by wildfires in a mining-impacted boreal landscape. Environ Res Lett 19(6):064024
59. Hoornweg D, Sugar L, Trejos Gómez CL (2011) Cities and greenhouse gas emissions: moving forward. Environ Urban 23(1):207–227
60. https://www.cer-rec.gc.ca/en/data-analysis/energy-markets/provincial-territorial-energy-profiles/provincial-territorial-energy-profiles-alberta.html. Accessed 23 Nov 2024
61. https://www.iedm.org/56509-how-do-canadian-provinces-fare-in-terms-of-ghg-emissions-per-capita/. Accessed 23 Nov 2024
62. Tian P, Zhong H, Chen X, Feng K, Sun L, Zhang N, Hubacek K et al (2024) Keeping the global consumption within the planetary boundaries. Nature 1–6
63. https://www150.statcan.gc.ca/t1/tbl1/en/tv.action?pid=1110007501. Accessed 23 Nov 2024
64. Wikipedia. Accessed 23 Nov 2024
65. https://www.ubs.com/us/en/wealth-management/insights/global-wealth-report.html. Accessed 23 Nov 2024

References

66. https://www.ctvnews.ca/business/some-of-canada-s-wealthiest-billionaires-according-to-forbes-1.6946497. Accessed 23 Nov 2024
67. Dhammapala R (2019) Analysis of fine particle pollution data measured at 29 US diplomatic posts worldwide. Atmos Environ 213:367–376
68. World Health Organization (WHO) (2016) Ambient air pollution: a global assessment of exposure and burden of disease
69. https://www.canada.ca/en/health-canada/services/publications/healthy-living/health-impacts-air-pollution-2021.html. Accessed 21 Nov 2024
70. Jha A, Nauze AL (2022) US Embassy air-quality tweets led to global health benefits. Proc Natl Acad Sci 119(44):e2201092119
71. Sheptycki J (2003) The governance of organised crime in Canada. Canad J Soc Cahiers Canadiens de Sociologie 28(4):489–516. https://doi.org/10.2307/3341839
72. Pellow DN (2004) Garbage wars: the struggle for environmental justice in Chicago. MIT Press
73. Germani AR, Pergolizzi A, Reganati F (2018) Eco-mafia and environmental crime in Italy: evidence from the organised trafficking of waste. In: Green crimes and dirty money. Routledge, pp 42–71
74. West G (2017) Scale the universal laws of life, growth, and death in organisms, cities, and companies. Penguin
75. Taylor et al (2012) Global urban analysis for 2010 and the world according to GaWC website for 2020. Accessed 6 June 2024

Open Access This chapter is licensed under the terms of the Creative Commons Attribution-NonCommercial-NoDerivatives 4.0 International License (http://creativecommons.org/licenses/by-nc-nd/4.0/), which permits any noncommercial use, sharing, distribution and reproduction in any medium or format, as long as you give appropriate credit to the original author(s) and the source, provide a link to the Creative Commons license and indicate if you modified the licensed material. You do not have permission under this license to share adapted material derived from this chapter or parts of it.

The images or other third party material in this chapter are included in the chapter's Creative Commons license, unless indicated otherwise in a credit line to the material. If material is not included in the chapter's Creative Commons license and your intended use is not permitted by statutory regulation or exceeds the permitted use, you will need to obtain permission directly from the copyright holder.

Chapter 6
Seven Generations of Canada's Cities: The Halftime Report

Over the last 100 years, Canadian cities were transformed by many inventions. Radios, televisions, personal computers, cell phones, and the Internet changed the way people interacted within and between cities. Electricity became the most important source of energy, and the electricity grid grew across North America, becoming the largest, and most complex machine ever built. Purveyors of electricity never pushed, seriously, to reduce demand, but rather promoted new appliances like toasters, microwave ovens, vacuum cleaners, and services such as escalators, street and traffic lights.

Per person electricity use in Canada more than tripled and is now one of the highest in the world (national average of 17 MWh per person in 2017; high of 21 MWh in Quebec) [1]. With a secure southern border, abundant hydro potential in the north, and lots of potential customers in the south, the first Canada-US transmission line started sending electricity south in 1909 [2]. Quebec, for example made more than $2.5 billion in sales of electricity in 2023. However, in 2023, British Columbia became a net-importer of electricity and Manitoba surpassed available hydro capacity, requiring 122 GWh of gas-fired electricity (and a lengthening waitlist of high-demand customers) [3]. Quebec is also scrambling to install an additional 150–200 TWh by 2050 (almost two times current capacity) [4]. Ontario is planning for a doubling in electricity demand by 2050 [5].

In the early 1920s, Canada became more than 50% urban [6]. In 2011, the country was more than 80% urbanized with the Atlantic Provinces, the least urbanized region, around 60% urban [7]. Canada is now a 'post-urban' nation where the relative population of urban and rural areas are largely balanced, and ebb and flow as determined by affordability and connectivity issues. Smaller, more rural, communities, like those in the USA [8], are expected to decline in population unless they can draw residents from larger cities and maintain up-to-date infrastructure. National and provincial governments will likely prioritize infrastructure improvements in the faster-growing cities, especially Calgary-Edmonton, Vancouver, Toronto Region, Ottawa-Gatineau, and Montreal, although political pressures will continue to elevate the concerns of smaller communities.

In 1920, Canada was emerging from WWI with a growing sense of pride, having played an important role in the war. Reintegrating returning soldiers was challenging, as they often wanted to return to cities rather than the farms where many came from. Their labor in the agricultural sector was sorely missed, although manufacturing jobs were growing quickly in urban areas. The world was also dealing with the pandemic (often referred to as the Spanish Flu) and returning soldiers had high rates of venereal disease [9].

The cities of Canada ushered in the "Roaring Twenties" with jazz and nightclubs (especially in Montreal, Canada's largest city at the time), less stringent alcohol prohibition than in the USA, declining unemployment, and a massive shift to a manufacturing economy. The 1920s brought growing prosperity and a strengthening sense of well-being. This ended abruptly on "Black Monday," October 28, 1929. By mid-November 1929, the Dow had lost almost half its value, and at its lowest point on July 8, 1932, it was 89% below its September 1929 peak (a level not reached again until November 1954) [10].

The US stock market crash had a severe impact on Canada, ushering in a shared Depression. Canada's Gross National Product dropped 40% between 1929 and 1939, unemployment reached 27% as Canadian exports shrank by half, and one-in-five Canadians became dependent on government relief [11]. Added to this, farms in the west faced severe drought [12].

The 1930s saw hardship around the world, political changes, growing protectionism, and rising autocracies. Canada's unemployment rate stayed above 12% until the start of WWII in 1939 [12].

Canada played a critical role and expended enormous blood and treasure in WWII. Some believe Canada first became a country of consequence on D-Day, June 6, 1944. Canadians plump with pride when the Dutch thank them every year for Canada's role in liberating them from the Nazis.

When looking out to 2120 and the second half of Canada's urban history, a few things become apparent. The rest of the world, having passed the 50% urban mark in 2008, will likely reach the fully urbanized levels (more than 80% urbanized) before 2120, except for parts of Sub-Saharan Africa. Global population will likely have peaked and be declining by 2120. And like Canada today, global populations will ebb and flow based on the demand for migrants, and the receptivity of regions where potential migrants want to resettle. Demand will be influenced by climate (likely substantively changed), relative quality of life, and economic opportunities. Receptivity to potential immigrants will be a blend of need, ability to accommodate, and local culture and openness.

Macrotrends will continue to buffet Canada's cities. These trends will be both 'home-grown' and global. In looking back and then forward four broad themes emerge: (i) demographics and economic productivity; (ii) climate change and other threats to planetary systems; (iii) energy and material flows; (iv) the changing role of cities and borders.

A key difference between the first half of Canada as an urban nation (1920–2020) and the second half (2020–2120) is the role of Canada within the global family of nations. Canada's population grew at a robust 2% annually with GDP growing even faster (2.7% annually). If Canada maintains its relative attractiveness to potential immigrants, population growth is likely to continue at around 1–1.5% annually. This may change as the world reaches peak population, likely around 2085, and then slowly declines for at least the 50 years after then; however, Canada is well-positioned to remain a key destination for immigrants and foreign workers.

Despite the comparatively high and continued population growth of Canada, the country's relative size and economic influence will continue to decline from the zenith reached around 1975.

Anecdotally, people sense a shift in the tenor of conversation, more stridency, less civility perhaps. This varies within and between countries. The 2024 World Happiness Survey observed a significant divergence in subjective well-being (SWB), or happiness, between Canadian age cohorts. Canada's old (60+) ranked their SWB 8th globally, one of the world's highest. However, Canada's young (< 30) ranked their SWB 58th globally: the largest difference among the 71 countries surveyed [13].

Canada may have reached the 50% urban level earlier than most countries; however, the country's collective thinking remained focused on developing the nation's abundant resources and safeguarding sovereign borders and global stature.

Canada seems to believe in an almost limitless supply of freshwater, trees, oil, gas and coal, fields of wheat, streams and lakes with beaver, fish and enormous hydropotential, farms, and minerals. The Soldier Settlement Act revised in 1919, highlighted this mindset. The Act crafted by Canada's political leaders gave 160 acres (65 ha) at little or no cost to returning soldiers to use 'idle lands' and develop rural areas [14].

Social unrest was thought to foment in cities. Labor activism such as the Winnipeg General Strike of 1919 and May Day celebrations in Toronto, Montreal, and other cities threatened agricultural development and business interests. The Veterans Land Act of 1942 again attempted to urge soldiers to take up farming upon return (this time after WWII and mostly on expropriated Indigenous lands, despite treaty obligations) [14]. Below Canada's surface however was an urban current drawing in returning soldiers and empowered women. The number of farmers steadily declined. In 1931, one-in-three Canadians worked in the agriculture sector. In 2021, this had declined to one-in-61 [15].

WWII also catalyzed infrastructure to open hinterlands and provide connections across the countryside. The Alaskan Highway, for example, was hastily built through British Columbia and Yukon Territory to provide a land route between the US mainland and Alaska. The Trans-Canada Highway, St. Lawrence Seaway, and the 63 stations of the Distant Early Warning (DEW) Line, were built in the 1950s to keep the nation open and safe for resource development and transport.

With national priorities and global imperatives, Canada's cities may have been less visible. However, as a strong current under the surface, Canada's urban growth drove much of the country's history for the last 100 years. For the next 100 years, Canada is likely to continue with two broad themes: (i) national directives and attempts by government to calm the waters of regional shifts and geopolitical turbulence and (ii) the 'care and feeding' of cities as urban areas continue to grow and provide disproportionate funds to national finances.

For the last 100 years, the relationship between the national government and Canada's cities and regions has been mutually supportive. The Canadian passport is top tier (ranked 8th in the world by Passport Index). Canada's macroeconomic environment has largely been positive, and exemplary among OECD-member countries. In 1950, Canada was the first major country to adopt floating exchange rates [16]. The country's current debt-to-GDP ratio is lowest among G7 countries.

Key trends and milestones of the last 100 years include:

- Population and demographic shifts (from 8.7 million in 1920 to almost 37 million in 2020); Canada reached the 50% urban level in the 1920s, while the world's overall urban population passed the 50% level in 2009.
- Relative to overall Canadian population, the share of Atlantic Canada, Manitoba, and Saskatchewan population roughly halved, while British Columbia and Alberta doubled.
- The age of abundance. In Canada per person energy use increased fivefold, material use tenfold, per capita GDP grew about 2.5% per year to $43,600 in 2020. Most of this growth occurred as part of the Great Acceleration starting around 1950.
- Canada emerged as one of the most energy- and materials-intensive societies in the world.
- Several connected global planetary boundaries crossed 'safe' limits, e.g., climate change, loss of biodiversity, land-use changes, and ocean acidification.
- Sovereigntist and French language pressures slowed growth in Quebec (significant impact on Montreal).
- Rise of the USA as global superpower.
- Friction intensified between many Canadian cities and upper levels of government on fiscal space and democratic representation.
- The start of reconciliation with Canada's Inuit, First Nations, and Metis (more than half of the country's "Status Indians" now live in cities)

- Geopolitical shifts ending colonialism, diminishing the influence of Europe and the emergence of Asia; about 120 new countries (out of today's total of 193) emerged in the last 100 years.
- Borders changed, globally growing in length several-fold, while becoming both less (e.g., USA, Canada, Mexico) and more permeable (e.g., within European Union) depending on location and what they are trying to restrain.

The most likely trends to impact Canadian cities over the next 100 years include (Tables 6.1 and 6.3):

- Continued, and accelerating population and demographic shifts.
- Regional agitation, e.g., eastern Canada's population decline (especially in Atlantic Canada) relative to increases in western Canada.
- Global peak population (likely around 10.6 billion in 2085); Canada's population could surpass 100 million (impact of immigration intensifies).
- Continued degradation of planetary systems, e.g., ocean acidification, loss of biodiversity.
- By 2120 global temperature increases likely exceed 3.0 °C, major tipping points crossed; more than 500 million climate migrants.
- Climate change (emphasis on decarbonization, need to adapt to a changing climate, geopolitical tensions over geoengineering, and agitation for loss and damages compensation).
- Canada's productivity uncertain (economic growth as measured by GDP)—trend is declining relative to the last 100 years.
- Continued tension with cities for greater fiscal and democratic representation, and concern over the debt loads of upper levels of government.
- Emergence of more durable 'urban–rural' and regional partnerships in Canada.
- Probable global shift toward circularity and sustainability (less energy- and material-intensive society, declining demand).
- Possible *black swan* events such as pandemics and natural disasters (e.g., earthquakes, volcanos, geomagnetic storms), likely crossing of several climate and other biophysical tipping points (Table 6.2).
- The increase in global migration forces more permeable borders and transient residency; critical role for Canada.
- Further erosion of public trust in political systems.

Key Issues for Canada

- Severity of global climate impacts and scale of resulting climate migrants and refugees wanting entry to Canada (relative to other countries).
- Beyond 2050 Canadian climate likely to experience significant increase in variability.
- Productivity of Canada's economy, overall well-being; related ability to attract and keep immigrants.
- Degree of social unrest and local and international agitation, especially in cities.
- Overall ecosystem health, in specific cities, e.g., heat islands, and regionally, e.g., Great Lakes water quality.

Table 6.1 Global events and trends

2030s
Global average temperature likely exceeds the 1.5 °C threshold. Primary energy from non-abated fossil fuels—81%. Coal, oil, and gas demand have all peaked, steep decline for coal to 2060, slower declines for oil and gas. Corporate space industry market capitalization exceeds $1 trillion (satellites in space that grew annually by 30% for more than a decade, along with much more space debris, require global cooperation, and regulation of businesses, to protect system integrity). China's political apogee. Fastest urbanizing regions: Sub-Saharan Africa, South Asia. "Other" surpasses French as mother tongue in Canada, English drops below 55%. India's economy surpasses Japan's. Manufacturing in space, e.g., lenses, drugs, fiber optics, quantum computing, and 3D-printing
2040s
Continued decline (population and relative share of global economy) of Russia, Germany, and Italy. Growing geopolitical influence of Turkey and Poland. Mexico's economy surpasses Canada. Fastest urbanizing regions: Sub-Saharan Africa, Central Asia. Widespread competition between Europe, USA, Canada, and other nations for immigrants. Total global space economy exceeds $2 trillion. More than half the earth's total landscape has undergone major change—state shift in earth's biosphere possible. International negotiations advance on geoengineering
2050s
'Battle stars' in space (mainly USA; military tensions). As many as 800 Mn people water-stressed in Africa (IPCC). Fastest urbanizing region: Sub-Saharan Africa and Latin America. Primary energy from non-abated fossil fuels—65%. Global material extraction 180 Gt per year (annual increase ~ 2.3%). Nigeria and (possibly) Pakistan surpass USA as world's third and fourth largest countries. Asia exceeds 50% of global GDP (North America and Europe drop from 52% in 2000 to 23% combined). Foreign-born and non-permanent resident population exceeds 40% of Canada's total population. "Net-zero" GHG emissions pledges largely unmet. More than half of Canada's economy generated by Montreal, Toronto, and Vancouver alone, almost an additional 20% from Calgary, Edmonton, and Ottawa-Gatineau. More than 20 advanced economies have an old-age dependency ratio above 50% (for every two working adults at least one person 65 years or older)
2060s
Possible solar-based power; US space supremacy. Rapid economic growth. Fastest urbanizing regions: Northern Europe, Canada. World fertility rate (2060) 2.06, high-income countries 1.65. India reaches peak population of 1.7 Bn (~2063) declines to 1.53 Bn in 2100 (greater decline may be possible through climate-induced emigration). Advances in robotics reduce emphasis on immigration. Variance of jet-stream in northern latitudes (attributed to global warming). Canada surpasses a 50% old-age dependency ratio. Population 'busts' common in many countries, possible restrictions on emigration, larger global cities experiencing steep population declines
2070s
Fastest urbanizing region: Arctic. Large-scale geoengineering. World fertility rate (2070) 1.98, high-income countries 1.65. Global average temperature likely exceeds the 2.0°C threshold. Climate tipping points of Greenland and West Antarctic ice sheets, and Boreal permafrost thaw underway
2080s
Mexico–US conflict possible. Global population peaks (likely around 10.4 Bn in 2086). World fertility rate (2080) 1.92, high-income countries 1.66. More than 3 billion people exposed to unprecedented heat (outside human climate niche, ≥ 29 °C). Loss of Barents Sea ice and decline of Labrador Sea SPG convection system. Loss of alpine glaciers and most coral reefs

(continued)

6.1 Demographics and Productivity

Table 6.1 (continued)

2090s	
World fertility rate (2090) 1.88, high-income countries 1.67. More than 5% ocean area suboxic (void of marine life). A global, collective, population goal likely established (with a future date for long-term population stability, i.e., target date for a 2.1 fertility rate)	
2100s	
World fertility rate (2100) 1.84, high-income countries 1.66. Global average temperature increase likely exceeds 2.9 °C. Canada's population likely to have peaked	
2110s	
World fertility rate (2110) 1.78. Japan's population declines below 50 million	

Compilation of references from: Friedman [57]; Altman [58]; Marshall [59]; Vince [60]; Khanna [24]; Bricker [61]; IPCC WG3 (2023); Clarke [62]; Lenton et al. [63]; Rhodium Climate Outlook (2023); Barnosky et al. [64]; Wei et al. [65]; WEF and McKinsey (2024); Shaffer et al. [66]; Armstrong McKay et al. [67]; Osman et al. [68]; Stanley [69]; UNEP 2014 *Emissions Gap Report* suggests a 3.1 °C global increase by 2100.

- Possible claims on Canadian sovereignty, geopolitical tensions, especially from shifting locus of economic activity—from OECD-member countries to Asia (East and South) and eventually Africa.
- Continued strength of USA and amicable tri-lateral relations, within USMCA framework or subsequent.
- Infrastructure and provisions of urban services *vis-a-vis* climate change, staffing capabilities, aging facilities, and maintenance.
- Global peak population, date of peak, rate of decline from the peak, differences across countries.
- Peak resource use (dates and magnitude), including oil and gas production post-2030.
- Resilience and climate adaptation, especially of major cities and large-scale infrastructure.
- Geopolitical tensions and shifts in energy and material supply chains.
- Impact from possible low-likelihood, high-impact events such as geomagnetic storms, earthquakes, pandemics.

6.1 Demographics and Productivity

Canadians will continue to help drive the global economy; however, Canada's collective global voice will be diminished relative to the last 100 years as the country's economy will be roughly half the size of the last 100 years; Canada's share of global population will drop from a high of 0.6% in the 1960s to less than 0.35% later this century. Canadians will also face challenges as their collective annual economic growth declines from almost 3% in the 1900s to less than 1% this century [17].

Table 6.2 Top 10 disruptions by likelihood and impact

Top 10 most likely disruptions	Top 10 highest impact disruptions
• People cannot distinguish true & false	• World war breaks out
• Biodiversity loss, ecosystem collapse	• Biodiversity loss, ecosystem collapse
• People cannot afford to live alone	• Healthcare systems collapse
• Biodata is widely monetized	• Civil war in the USA
• Billionaires run the world	• Emergency response is overwhelmed
• Downward social mobility is the norm	• Basic needs go unmet
• Emergency response is overwhelmed	• Cyberattacks disable infrastructure
• Mental health crises	• People cannot distinguish true & false
• Cyberattacks disable infrastructure	• Democratic systems breakdown
• Artificial intelligence runs wild	• Vital natural resources are scarce

From: Government of Canada. Disruptions on the Horizon, 2024 Report

Canada's low productivity, relative to other high-income countries, and high household, national and provincial debt levels, will limit city-based initiatives and transitions. The energy transition—the plan to decarbonize by 2050—will also require significant investments and socio-economic changes in urban areas.

During the next 100 years, one of Canada's (and North America's) fastest-growing cities, Toronto Region, will increase in population to 18 million, but will still likely drop from the world's 40th largest city (~ 2010) to not being in the Top 100. Montreal and Vancouver may both drop out of the Top 200 largest-cities list. Like the USA, Canada's smaller secondary cities are likely to see widespread and significant depopulation [8].

6.2 Climate Change

Looking back 100 years on the climate system, humans released about 2,500 billion tons of CO_2e; two-thirds of that was in the last 35 years. Annual CO_2 emissions have increased by 50% since 1990 [18]. In 2023, a record 57.1 Gt CO_2e was released, a 1.3% increase over 2022 [19]. Previous GHG emissions have already caused significant temperature increases. July 2024, for example, was 1.21 °C above the twentieth century average [20]. Temperatures that were increasing by about 0.18 °C per decade (1970–2000) have now accelerated to an increase of at least 0.3 °C per decade [21].

Looking forward, there is considerable uncertainty associated with climate impacts (intensity and speed) and changes to the atmosphere, and related earth systems such as biosphere, cryosphere, and hydrosphere. However, a few trends are clear. The rate of GHG emissions is still increasing; net-zero by 2050 is extremely unlikely. The target is still important; however, global average temperature increases of more than 2.5 °C should be anticipated. This will likely trigger nonlinear climate tipping points, although dates are not certain. A scramble for geoengineering 'solutions' should also be anticipated [22].

Table 6.3 Trends in urban Canada

Demography • Declining populations • Aging population • Increased life expectancy • Changes to family dynamics **Geopolitical Tensions** • Supply chain disruptions • Declining globalization; bifurcating world into BRICS and "the West" • Struggle for power in the digital world and space • Climate mitigation efforts delayed, energy security emphasis **Climate change** • Climate mitigation targets slipping • Challenges of adaptation • Major tipping points crossed • Climate a precursor to other ecosystem threats, *e.g.*, biodiversity loss **Economic shifts** • New industries, service providers • Changing jobs, type, and location • Macroeconomic instability, declining productivity, and growth • Circular and sharing economy • Increasing inflation and macro-economic uncertainty **Urbanization** • Changing urban populations, in Canada and globally • Housing shortage • Inadequate mobility, significant congestion • Growing influence of African cities, post-2040 **Inequality** • Reduced access to social services • Affordability • Increasing gap between rich and poor, by city, region, globally **Governance** • Increased populism • Impact of miss-inadequate-information • Loss of civility and trust • Foreign interference in political process	**Energy and materials transition** • Decarbonizing electricity; increased demand and grid development • Infrastructure needs; greater climate hardening • Material supply, e.g., rare earth and critical minerals • Energy from space and/or fusion—post-2060 • Peak (solid) waste (~ 2085) • Energy transition (*e.g.*, "net-zero" by 2050) delayed, beyond 2060 **Digital and technology transformation** • Development of artificial general intelligence • Digital security, ownership, and availability • Space systems development (largely military applications) • Geoengineering • Data systems—municipal service **Migration and immigration** • Climate migration (refugees and economic migrants) • Competition for human capital, attracting immigrants • Integration (and retention) of immigrants • Increased efforts to *strengthen* borders 2025–2040; service provision concerns from some municipalities • More borders, changing permeability **Infrastructure** • Climate hardening • Financing, new approaches • Labor, trades, and supply chains • Declining level of state of good repair • Biomimicry, nature-based systems • Criticality of regenerative urban systems (beyond resilience) **Changes in health care and education** • Aging population • Next pandemic • New healthcare technologies, *e.g.*, robotics, AI • Diseases of the aged; declining mental health • Online education; role of AGI • Increased foreign students

Climate impacts and increases in carbon, methane, and nitrogen are reducing the buffering capacity of planetary systems. Waterbodies may shift quickly to an anoxic state, invasive species may move more aggressively into altered ecosystems, and climate variability may increase markedly. Cities will need to be nimbler as they respond to these impacts. Resilience and adaptation are among the most important attributes of successful cities.

Lenton et al. suggest that by end of century (2080–2100), current policies are leading to around 2.7 °C global warming and could leave one-third of people outside the 'human climate niche'. In a world with 9.5 billion people, more than three billion would be in areas where the mean annual temperature exceeds safe limits. At least 28 countries would have 10 million or more people exposed to unprecedented heat.[1] India, Nigeria, and Indonesia alone will have more than one billion people in unsafe areas.

In addition to heat, extreme water stress already affects 25 countries—home to a quarter of the world's population. At least 50% of the world's population—around four billion people—live under highly water-stressed conditions for at least one month of the year [23].

This large potential source of migrants moving away from areas of rising temperature will be supplemented with people living in increasingly water-stressed areas with associated food insecurity, as well as people in harm's way from rising sea levels, increasing windstorms, and natural disasters such as earthquakes and tsunamis. Wars and acts of aggression, which could intensify in an increasingly climate stressed world, would also likely continue to be a major source of migrants. By the end of the century a third to a half of humanity may be moving away from harm's way [24].

In a harbinger of things to come, West Africa's February 2024 humid heat was made 10 times more likely by climate change estimates World Weather Attribution [25]. Climate change is already adding 4 °C to some regional temperature increases leading to average 'danger' levels of Heat Index values of about 50 °C. Locally, values even entered the level of 'extreme danger' with values up to 60 °C. As these increases become more common and attribution to current and historic GHG emissions clearer, high-income countries, with relatively high GHG emissions, should anticipate escalating pressures for reparations (and more mitigation).

Although Canada's climate challenges are likely to be less severe than many other parts of the world, significant changes should still be anticipated. Increased forest fires; degraded water quality, especially in smaller lakes; higher intensity rainfall and associated flooding; strained electricity transmission grids; and heat stress on infrastructure and cities are all likely.

---

[1] Top countries with populations exposed to unprecedented heat with 2.7 °C warming above pre-industrial levels (average annual temperature $\geq$ 29 °C): India (617.7 million); Nigeria (323.4); Indonesia (95.2); Philippines (85.6); Pakistan (84.1); Sudan (79.5); Niger (72); Thailand (54.1); Saudi Arabia (48.9); Burkina Faso (47.2); Mali (43.1); Ghana (32.7); Vietnam (32.3); Chad (29.4); Malaysia (27.7); Myanmar (26.5); Benin (22.6); Yemen (19.3); Columbia (18); Brazil (17.1); United Arab Emirates (16.9); Cambodia (16.8); Senegal (16.5); Venezuela (15.5); Cote d'Ivoire (13.9); Sri Lanka (13.7); Somalia (11.2); Ethiopia (10.2).

Swiss Re reports that under the current trajectory, global average GDP could be 11%–14% less by mid-century. Impacts are expected to be most severe in the ASEAN region (29%) and lowest in North America (8%) [26]. This disparity in climate impacts (somewhat opposite of GHG emissions) will foment global social unrest.

Canada's temperate climate and relatively plentiful freshwater enhances the region's desirability for the rest of this century. Temperature increases and associated population shifts will likely place significant immigration (and refugee) pressures on Canada. Canada may be under pressure to offset its disproportionate contribution to climate change by being more welcoming to climate migrants.

6.3 Energy and Material Flows

The World Bank estimates that three billion tons of metals and minerals are needed to achieve a below 2 °C future [27]. This is largely to provide what the IEA argues is required to meet net-zero emissions by 2050: no sales of new internal combustion engines after 2035, and the global electricity sector must reach net-zero emissions by 2040 [28].

The Great Acceleration beginning in the 1950s saw Canada's urban metabolism grow to be among the most energy- and materials-intensive in the world. Canada grew quickly to be one of the most wasteful societies in the world, as well as a key provider, and champion of similar high materials use internationally. Canada played an important role in the last 100 years in mining and resource development. How this continues over the next 100 years is not yet clear (see Chap. 5).

During a January 2024 discussion between Michael Liebreich and Mark Cutifani, Chairman of Vale Base Metals, the 3.6 GW high-voltage direct current interconnector to carry solar and wind-generated electricity from Morocco to the UK was discussed. Host Liebreich mentioned that he had learned on an earlier podcast with Simon Morrish, the CEO of Xlinks, that due to capital constraints ($34 billion) the cable was going to be made of aluminum, rather than copper which is better conducting but more expensive. Chairman Cutifani opined that perhaps a new financing approach is needed where a company like Vale might lease the copper, rather than sell it outright [29].

Sustainability and its derivative sustainable development may be negotiated through countries, although the process as structured through the UN is problematic. Consensus is required, and with more than 75 of the 193 countries having populations under five million, historic grievances, and some countries and regional factions blocking progress, and some even having disdain for the process, agreements are often difficult. Pursuing and revising the sustainable development goals will likely be delegated (or assumed) by urban areas as the key attributes such as energy and material flows, GHG emissions, equity, and health care, are more typically observed at the local government level.

6.4 COVID-19

Arundhati Roy reflecting on the COVID-19 pandemic provided a powerful essay 'The pandemic is a portal' (Financial Times, April 2020). She wrote:

> Historically, pandemics have forced humans to break with the past and imagine their world anew. This one is no different. It is a portal, a gateway between one world and the next. [30]

Arguably the main event at the 'halftime show' for Canada's cities was COVID-19. COVID-19 exacerbated existing trends, along with highlighting many of the current strengths and weaknesses of cities. COVID-19 will serve as a social stratigraphic marker for cities. Post-COVID life in Canada's cities has already emerged with more work from home and hybrid work. The importance of essential workers and supply chains was highlighted during COVID. Governments will attempt to buttress these potential vulnerabilities, although options are not always readily available.

The importance of relatively small inputs. For example, vaccines, PPE, and computer chips in new-car manufacturing. Vaccines also provided a glimpse into social collectives and the growing trend of societies to fracture along ideologies as tribes.

The trucker convoy, protesting mostly against vaccine mandates, that occupied Ottawa for more than three weeks in winter 2022 also highlighted critical urban issues. The subsequent Rouleau Report illustrated the poor coordination between levels of government [31]. This may be particularly prevalent in Ottawa with a larger federal government role, and a more hands-off province of Ontario government. The debate over vaccine mandates was largely fueled by ideologies, and not driven by a pragmatic city-management perspective.

The precipitous drop in transit ridership during COVID illustrated a vulnerability when transit costs are borne mostly through fares. For example, Toronto transit is funded 70% from rider fares, while rider fare contributions in comparator cities of Chicago, Boston and Houston, make up only 40, 33, and 13%, respectively (the rest from state and national government support) [32].

On this side of the COVID portal, as the hinge moment unfolds for Canadian cities, the change is increasingly clear. The energy transition has started and is serving as an early chapter in the transition to greater sustainability. Global populations are passing a transition point as well. Still some time away, but stabilization and eventual decline are clear (likely before 2085).

As some city folk retreated to their cottages during COVID, they often raised the ire of locals, who argued that the new residents overwhelmed local infrastructure such as health services. However, a precedent was set, and with access to improved online meeting platforms and better connectivity, employers are more open to have workers work remotely. Flexible working arrangements are emerging. These will be helpful as more fluid residency and employment arrangements will be needed for temporary foreign workers and climate migrants. The flexible work arrangements can also reduce congestion and transportation emissions.

Post-COVID, a growing number of smaller municipalities in Canada are adopting four-day work weeks to attract and retain more talent. The Township of Algonquin

Highlands, Ontario for example, announced in May 2024 that it would permanently move to a "compressed" work week. Instead of working five days per week and eight hours per day, municipal employees will work 10-h days four days a week [33].

Surveys from USA, Europe, and Asia suggest that hybrid work provides equivalent compensation to employees of an 8% increase in salary. And preliminary analysis from the USA shows couples may be open to about 0.3–0.5 more children when both work from home one day or more a week [34].

Another issue highlighted by COVID is the power and risk associated with seemingly small things; sensitivity in the supply chain to items like personal protective equipment (PPE) or ventilators. Residents in larger cities who live in smaller condominiums and apartments, also exhibited a strong need to be able to regularly venture outside, even if just for a few minutes.

6.5 State of Not-so-Good Repair

With Infrastructure Canada, Statistics Canada regularly provides Canada's Core Public Infrastructure Survey (last reference year 2020). The review is an extensive assessment of public transit, roads, bridges, tunnels, stormwater, wastewater, potable water; culture, recreation and sports facilities; public social and affordable housing; solid waste and asset management. More than 90% of the infrastructure monitored is owned by municipalities. In all asset classes, overall infrastructure quality declined between 2018 and 2020. About 40% of infrastructure is rated very poor, poor, or fair.

A significant portion of linear water infrastructure was over 50 years old in 2020. Close to one-in-five kilometers of water, sewer, and stormwater pipes (86,533 km out of 472,488) was reaching the end of its useful life, having been built prior to 1970 [35]. Recent highly disruptive water main breaks in Calgary and Montreal support the finding of declining quality in water infrastructure. Canada's aging and inadequately maintained infrastructure is now meeting a changing climate, and population growth rates exceeding 2% per year.

6.6 London Burns

On a breezy September 2, 1666, after a warm summer, just past midnight, a fire started at a bakery in Pudding Lane, London. Lord Mayor, Sir Thomas Bloodworth's indecisiveness delayed demolitions for a firebreak. By the time he finally agreed to the order late that Sunday night, the fire was raging unchecked. Civility in the streets broke down as rumors arose of suspicious foreigners setting the fires. Catholics, as well as the French and Dutch, England's enemies at the time, were blamed and became victims of street violence. The conflagration lasted four days and destroyed more than 80% of London [36].

The fire gave rise to building code revisions (exteriors of buildings had to be brick or stone), wider streets and more open access to the River Thames, as well as the introduction of professional fire brigades. The fire destroyed more than 13,000 homes at a time when insurance did not exist. Physician Nicholas Barbon capitalized on the business opportunity, setting up the first insurance company in 1667. Further insurance companies were set up, including the Sun Fire Office, which was established in 1710 and is now the oldest insurance company in the world [37].

After the Great London Fire, King Charles II encouraged the homeless to move away from the city and settle elsewhere, immediately issuing a proclamation that "all Cities and Towns whatsoever shall without any contradiction receive the said distressed persons and permit them the free exercise of their manual trades" [36].

More than 200 years later after a long drought, strong winds from the southwest quickly spread a fire that started in or around a small barn belonging to the O'Leary family in southwest Chicago. From October 8–10, 1871, the conflagration killed approximately 300 people, destroyed roughly 9 km^2 of the city including over 17,000 structures, and left more than 100,000 residents homeless. The rapid destruction of the water pumping system contributed to the extensive damage of the mainly wooden structures [36].

Fires such as those in London and Chicago gave rise to advances in building codes, establishment of fire departments, and fire hydrants. Fire services now consume about 7% of Canadian municipal budgets.[2]

A warmer, often windier and drier, climate is however changing fire management services in cities. Lahaina, Hawaii; Valparaiso, Chile; Paradise, California; Athens, Greece—along with Slave Lake in 2011, Fort McMurray in 2016, Lytton, B.C., in 2021, Halifax in 2023, and Jasper in 2024—fires in these cities are a harbinger. Some argue we have entered the Pyrocene [39] where fires from outside the city are as great a threat as fires originating from within the city.

In 2023, 230,000 people in Canada had to be evacuated for wildland fires [40]. In the USA, wildland fires resulting from transmission line failures led to financial impacts and even bankruptcy for power utilities such as Pacific Gas & Electric, Hawaiian Electric, and Xcel Energy. Transmission lines and grid reliability are susceptible to more intense fires. Traditional municipal fire departments are not yet equipped to deal with these fires. Nor the increasingly common flooding, ice storms, and heat events, that are accompanying a changing climate.

First responders, municipalities, community service agencies are adjusting to higher numbers of internally displaced persons (IDPs). Remote indigenous communities are disproportionately affected as they experience higher rates of evacuations and threats.

---

[2] Based on Ontario values from Found [38].

6.7 Follow the Trends

Looking back 100 years, the trends that impacted Canadian cities are readily apparent. Putting yourself in the policy advisor of 1920 some would have been easy to predict and some less so. Demographics and national rates of immigration are one of the most powerful trends influencing Canada's cities. The inexorable trend of urbanization and population declines (mainly through fewer children per female) impacted Canadian cities throughout the twentieth century. Canada enjoyed a baby boom as large as any country; however since 1964, this boom has had a few modest echoes while steadily declining.

Political trends would have been less evident in 1920. For example, Montreal was more cosmopolitan and larger than Toronto. Winnipeg was larger than Vancouver and Quebec City larger than both Calgary and Edmonton. Changes in relative size of Canadian cities are mainly driven by responses to politics and public policy. For example, Montreal's (and Quebec City's) relative decline is linked with Quebec separatism and language initiatives.

For the next 100 years census metropolitan areas (Statistics Canada definitions) take on an even greater role. Some of the smaller CMAs such as Kelowna and Halifax may continue to be the fastest-growing, but the main urban areas of Montreal, Toronto, Ottawa-Gatineau, Calgary-Edmonton, and Vancouver, plus Quebec City, Halifax, and Winnipeg, will determine most of Canada's economy. These communities will welcome most immigrants and together will drive the national economy.

The region to experience the most change will be northern Canada. Northern shipping routes will likely open, cities are expected to be among the fastest-growing in Canada. Development in this region will need to be particularly sensitive to environmental concerns, e.g., melting permafrost and shifting winter ice.

6.8 One Number to Rule Them All: Looking Beyond GDP

The modern concept of GDP was developed by Simon Kuznets in a 1934 US Congress report trying to capture the relative strength of various countries' economies. The work built on UK's efforts in the 1920s and 1930s that calculated national income and expenditures on a quarterly basis [41]. Kuznets warned against using GDP as a measure of welfare. In 1959, economist Moses Abramovitz further cautioned "we must be highly skeptical of the view that long-term changes in the rate of growth of welfare can be gauged even roughly from changes in the rate of growth of output." Despite these warnings, during the 1950s and 1960s, GDP as a measure of economic strength became widespread.

President Kennedy felt the need to echo the warning.

Too much and for too long, we seemed to have surrendered personal excellence and community values in the mere accumulation of material things that Gross National Product[3] counts air pollution and cigarette advertising, and ambulances to clear our highways of carnage. It counts special locks for our doors and the jails for the people who break them. It counts the destruction of the redwood and the loss of our natural wonder in chaotic sprawl. It counts napalm and counts nuclear warheads and armored cars for the police to fight the riots in our cities.

Yet the gross national product does not allow for the health of our children, the quality of their education or the joy of their play. It does not include the beauty of our poetry or the strength of our marriages, the intelligence of our public debate or the integrity of our public officials. It measures neither our wit nor our courage, neither our wisdom nor our learning, neither our compassion nor our devotion to our country; it measures everything in short, except that which makes life worthwhile. And it can tell us everything about America except why we are proud that we are Americans.

—John F. Kennedy's remarks at the University of Kansas, March 18, 1968

The Bretton Woods institutions established in 1944 (the International Monetary Fund, IMF and the World Bank, WB) solidified the use of GDP as a tool for measuring a country's economy. Lending volumes (and interest rates), and differentiated fees, and capital contributions were often defined by a country's GDP. The World Bank published every country's GDP in its Annual Report and GDP as a measure of economic output became common over the last half of the twentieth century. The power of the metric grew as governments, policy advisors, and researchers relied on its annual publication.

The Cold War solidified GDP as a critical metric. In the absence of battlefields where adversaries clashed and the winning side could be easily monitored, economic prowess took on greater importance. Growth mattered to let citizens know which side was winning [42].

Today, GDP—as a measure of economic output—is one of the most powerful metrics driving political processes. Recognizing the limitations of GDP as a measure of a country's progress, other metrics such as the human development index [43] (HDI) were introduced. King Jigme Singye Wangchuck of Bhutan believing GDP was such a poor metric of his country's overall performance decided instead to report Gross National Happiness. Suffice to say GDP is a blunt and simple metric as an indicator of a country or community's overall health and prosperity.

However, like how taking a temperature provides the healthcare worker a quick glimpse into the patient's health, tracking an economy's GDP provides a useful, albeit highly limited, measure. GDP became such a powerful metric because it is relatively straightforward to measure, is comparable over time and across countries and regions, and by publishing values annually, or even quarterly in some cases, the metric influences behavior and public policy. SDG 17.19 calls for enhancements to GDP as a measure of well-being (and national statistical capacity to measure

---

[3] GDP and GNP (domestic vs national) are often used inter-changeably. GNP is based on GDP with the country's global net incomes and outflows. Measured production by a country's citizens at home and abroad.

progress)[4] [43]. In developing these complementary metrics, a fundamental challenge arises in that reaching consensus among 193 member countries is difficult but is especially so when these countries may have voice for their territorial progress, but usually not agency for its achievement (or certainly not complete agency).

GDP, or Gross National Product that attempts to measure a country's economy even if some of these activities take place outside territorial limits, is a simple measure of a country's relative economic size. GDP was a wartime concept that helped a country's leaders know how much financial might could be directed toward military activities, and how potential rivals stacked up against each other. GDP became an ubiquitous metric in the twentieth century as the number of countries rapidly expanded (almost tripling) and an easy rule of thumb was needed to place countries in an economic hierarchy. GDP is a bit like measuring the width of moose antlers to size up the likely victor in the forthcoming fight. The NATO-target to spend 2% of GDP on national defense is another example of the power of GDP as a metric.

In 2009, the French government released a report co-authored by Nobel Prize-winning economist, and former World Bank Chief Economist, Joseph Stiglitz that called for an end to "GDP fetishism". A year later British Prime Minister, David Cameron's government announced that it would begin to survey happiness in addition to other economic measures.

6.9 Looking Beyond GDP

In the March 2024 issue of the IMF's flagship Finance & Development publication, the *World Happiness Report* was highlighted to lead the quest for more holistic measures of well-being [44]. The quality and reliability of the World Happiness Report, now published annually with broad support and an editorial board, provide assurances to governments, institutions, and academics to consider subjective well-being (happiness) as a credible metric for societal progress.

The field of subjective well-being (SWB), with its long history and international acclaim, was largely led by John Helliwell at University of British Columbia's Vancouver School of Economics [45].

The World Bank presented *City Indicators: Now to Nanjing* [46] at UN-Habitat's Third World Urban Forum, in Vancouver, June 22, 2006. The paper suggested the development of standardized indicators that are replicable, potentially predictive, and consistent and comparable over time and across cities. The World Bank supported the establishment of the Global City Indicators Facility and eventual World Council on City Data at the University of Toronto. The city indicators are now published as ISO standard 37,120.

---

[4] SDG 17.19: by 2030, build on existing initiatives to develop measurements of progress on sustainable development that complement GDP, and support statistical capacity building in developing countries.

The World Bank city indicators included SWB with an assessment provided for Toronto, Sao Paulo, and Belo Horizonte. Noticeable differences in the level of happiness between Sao Paulo and Belo Horizonte were observed, suggesting as granular a scale as possible. This finding was supported in a 2014 Statistics Canada review of SWB across Canadian neighborhoods [47]. A challenge in developing any metric at a country scale is the heterogeneity across a country and even as the Statistics Canada report found between neighborhoods in the same city.

6.10 The Difficulty in Measuring the 'Ease of Doing Business'

The World Bank's most read report was consistently the annual *Ease of Doing Business* report (2003–2021). Using various indicators, countries were ranked for business competitiveness. Many countries and agencies took the report very seriously; in some cases, the country's relative performance in the rankings was used to evaluate government officials and departments. The report was discontinued in 2021 over allegations of data manipulation. A challenge also arose as countries were evaluated in only one city (two in the USA; New York and Los Angeles, and eventually two in China: Beijing and Shanghai).

The 2019 World Bank *Ease of Doing Business* report ranked Canada the 22nd most competitive business economy in the world. The evaluation was made through ten categories such as starting a business, dealing with construction permits, getting electricity, contractual, and legal aspects. The categories were assessed through collection of some 285 data points. Activities fell under the mandates of the Government of Canada (~ 160 data points), Province of Ontario (~ 58), City of Toronto (~ 45), Toronto Hydro (~ 15) and professionals and companies (~ 8). Results could vary by city and by province (and electric utility), yet the values are presented as a single metric for Canada's overall ease of doing business score. This highlights the challenges of measuring something as complex and temporal as starting and running a business [48].

In devising a new or complementary metric for GDP, caution is needed to ensure *what* is being measured, and *where* it is being measured is understood. The prominence of GDP as a metric last century was driven by increasing numbers of countries (more than 120 new countries emerged in the twentieth century) and the influential role that agencies such as the United Nations and Bretton Woods institutions assumed. These organizations are managed by, and report to, countries. This helped to increase the overall stature of countries (relative to cities and regions).

The ease of doing business in Calgary may be as different between Toronto as it is with Los Angeles. Or the happiness of people living in Vancouver may be as different as those living in Ottawa as those living in Sao Paulo. The next 100 years will see improved metrics for overall sustainability, or flourishing, within major cities. This makes sense, as data systems and artificial intelligence tools, make refinement and targeting of information possible. This is also important as urban areas are where the lion's share of the economy is derived, and most people live.

As a minimum, in addition to disaggregated GDP, Statistics Canada should publish annual sustainability metrics for Canada's 10 largest census metropolitan areas (CMAs—See Chap. 2). Much of this is already in place as the Government of Canada, and Canadian institutions lead in discussions on future metrics of well-being [49].

6.11 The Next Summit

In the summer of 1944 representatives from 44 countries came together at the Bretton Woods resort in New Hampshire. Canada was there; in keeping with its size and global stature it mostly played a supporting role. Two key players were the UK and the USA, but the Soviet Union and China, as well as India and Brazil, were also influential [50]. Russia (as the successor state to the Soviet Union) did not join the IMF and World Bank until 1992.

The UK, impoverished after WWI and WWII, was represented by John Maynard Keynes. The USA on the ascendancy as the world's most powerful nation was led by the lesser known, but even more tenacious Harry Dexter White.

The USA was magnanimous in its support, but not without self-interest. A peaceful institution-based world would benefit the USA considerably. Canada championed a multistakeholder approach and has served as one of the world's loudest cheerleaders for a rules-based world order. This included global policing efforts like UN peacekeeping.

OECD-member countries influenced the rules and benefited significantly (the OECD was created in 1961). GDP as a key metric for a nation's 'success' was supported by the new institutions.

Wars still flourished and the Cold War (1945–1991) took decades to decide, and even then, it was not fully agreed-to as Russia's 2022 invasion of Ukraine suggests.

The Bretton Woods Summit was mostly about establishing monetary and trade structures. The 60 years of relatively stable trade that followed, especially benefitted the 'West' (OECD-member countries). Much of the wake of Bretton Woods was defining how the world would grow and share wealth. Global energy use increased from 28,000 TWh in 1945 to 144,000 TWh in 2005 (~ 400% increase) [51]. Global average wealth increased from $3360 per capita in 1950 to $9904 per capita in 2000 (~ 300%) [52]; while in the US GDP per capita rose from $2845 in 1960 to $44,123 in 2005 (~ 1400%) [53]. Global municipal solid waste volumes increased from about 187 million tons in 1950 to more than 1.3 billion tons in 2000.[5]

---

[5] Author estimate: 1950 urban population of 746 million with average per capita GDP of $3050, and annual average MSW rate 250 kg/year (187 million tons); 2000 urban populations of 2.96 billion, with per capita GDP of $6500, and annual MSW rate of 450 kg/year (1.3 billion tons).

OECD-member countries saw decades where GDP grew annually by 3–4%. More than two-thirds, the GHG emissions and solid waste (materials consumption) were in OECD-member countries. Much of Sub-Saharan Africa and countries like Brazil, Russia, India, and China were largely outside the economic mainstream. Cities within these countries operated under the macroeconomic environment provided by their national governments. Global average material consumption was much lower than OECD-member countries; however as illustrated in Chap. 3, China surpassed the USA as world's largest GHG emitter in 2006. India will likely surpass the USA in total GHG emissions before 2050.

The precursors to WWII, declining global trade, rising nationalism and polarization, are rising again. As pundits like Jeff Rubin opine, we are likely at war already [54]. The USA, Europe, Japan, and Canada on one side. Countries like Russia and China on the other side and many countries are still on the sidelines.

"Friendshoring" is proposed as a political response to global tensions—friendlier but with a much smaller group of friends. Canadian cities did not have much of a say in this, and there may well-be more visceral competition between Mississauga and Toronto than say between Toronto Region and Beijing. However, unlike Bretton Woods, the next summit would likely be more about access to critical materials, agreeing to an allocated share of earth system capacity (e.g., GHG emissions, biodiversity impacts), loss and damages compensation, data flows and security, and establishing proposed approaches to the movement of climate (and other disaster) migrants.

With 59.1% of voting shares in the IMF held by countries representing 13.7% of the world's population the IMF must urgently reform its constitution [55]. India and China, the world's two most populous countries, with 36% of the world's population, have a combined share of only 9%. Africa, the world's fastest-growing region, is even less-well represented.

Several researchers make the case for a clean energy Marshall Plan [56]. Modeled after the US assistance to Europe after WWII, a clean energy Marshall Plan, could take advantage of Washington's unique capacity to support a transition to low-carbon energy and more resilient materials access. The USA may wish to broaden the mandate to sustainability rather than only clean energy and may wish to partner with other countries to support financing and management capacities. In developing a new Marshall Plan, countries like the USA, Germany, Japan, UK, Australia, and Canada may want to establish partnership support mechanisms with a few key cities. A clean energy (or sustainability) Marshall Plan requires more than finance. Access to training and education, management capacities, and possibly migrant nodes are also needed. Key cities could anchor some of these programs on a long-term basis.

The 2024 election results in the USA (and other countries) suggest a need to structure this new global sustainability plan in a manner that facilitates a straightforward understanding by US citizens on how they, and their children, would benefit from the approach. The original Marshall Plan provide a compelling case for international engagement; however, many countries are experiencing growing nationalism, and angst about conditions at home.

References

1. https://utorontopress.com/blog/2019/01/17/canadas-shifting-energy-sources-a-comparison-with-eight-european-countries-1870-2000/
2. https://www.electricity.ca/knowledge-centre/about-the-electricity-sector/history-of-electricity/
3. Narwhal Magazine (March 28, 2024) The demand for power might make one of Canada's cleanest grids dirtier
4. Hydro Quebec (2023) Action plan 2035: towards a decarbonized and prosperous Quebec
5. Ministry of Energy (2023) Powering Ontario's growth
6. Goheen PG (1980) Some aspects of Canadian urbanization from 1850 to 1921. Urb Hist Rev 77–84
7. Statistics Canada (2014) Canada's rural population declining since 1851. Canadian Demography at a Glance
8. Sutradhar U, Spearing L, Derrible S (2024) Depopulation and associated challenges for US cities by 2100. Nat Cities 1(1):51–61
9. Suggit B (2017) Venereal disease in the first world war
10. Federalreservehistory.org. 17 Aug 2024
11. Chartered Professional Accountants Canada (2024) Financial crises roundup: a history of the biggest market shocks
12. Canadian Encyclopedia (2017) The Great Crash of 1929 in Canada
13. World Happiness Report (2024) Available at https://worldhappiness.report/ed/2024/happiness-of-the-younger-the-older-and-those-in-between/. Accessed 15 June 2024
14. Solomon L (2007) Toronto sprawls: a history. University of Toronto Press
15. Statistics Canada (2021) Canada's farm populations
16. Thiessen G (2000) Remarks by the Governor of the Bank of Canada. Montreal 4
17. OECD (2024) OECD economic outlook, interim report February 2024: strengthening the foundations for growth. OECD Publishing, Paris. https://doi.org/10.1787/0fd73462-en
18. US EPA (2023) Climate change indicators: greenhouse gases
19. UNEP (2024) The emissions gap report
20. NOAA (2024) National centers for environmental information (ncei.noaa.gov)
21. Samset BH, Zhou C, Fuglestvedt JS et al (2023) Steady global surface warming from 1973 to 2022 but increased warming rate after 1990. Commun Earth Environ 4:400. https://doi.org/10.1038/s43247-023-01061-4
22. Dyer G (2024) Intervention earth: life-saving ideas from the world's climate engineers. Random House Canada
23. WRI (2024) Aqueduct water risk atlas
24. Khanna P (2021) Move: how mass migration will reshape the world—and what it means for you. Orion Books
25. https://www.worldweatherattribution.org/dangerous-humid-heat-in-southern-west-africa-about-4c-hotter-due-to-climate-change/
26. Swiss Re Institute (2021) The economics of climate change: no action not an option, 34p
27. World Bank (2020) Minerals for climate action: the mineral intensity of the clean energy transition
28. International Energy Agency (2021) Net zero by 2050: a roadmap for the global energy sector
29. https://cleaning-up-leadership-in-the-age-of-climate-change.simplecast.com/episodes/ep151-mark-cutifani-7ZbcewHr
30. Roy A (2020) The pandemic is portal. Financial Times, 3 Apr 2020
31. Rouleau PS (2023) Report of the public inquiry into the 2022 Public Order Emergency
32. CodeRedTO (2018) Mixed signals: Toronto transit in a North American context
33. https://www.ctvnews.ca/business/rural-municipalities-in-canada-lead-the-way-in-4-day-work-weeks-to-combat-high-turnover-rates-1.6304194
34. Bloom N (2024) Finance & development. International Monetary Fund

35. Government of Canada. https://housing-infrastructure.canada.ca/plan/ccpi-ipec-eng.html. 18 Aug 2024
36. Wikipedia. 18 Aug 2024
37. BBC (2016) Five ways the great fire changed London
38. Found A (2012) Economies of scale in fire and police services in Ontario. In: IMFG working paper No. 12
39. Pyne S (2022) The pyrocene: how we created an age of fire, and what happens next
40. Natural Resources Canada (2024)
41. Coyle D (2015) GDP: a brief but affectionate history. Princeton University Press
42. Susskind D (2024) Growth: a history and a reckoning. Allen Lane
43. Dickson E (2011) HDI was launched in 1990 by UN economist Mahbub ul-Haq after requesting Amartya Sen to create "an index as vulgar as GDP but more relevant to our own lives"
44. Stanley A (2024) International monetary fund. Looking beyond GDP. Finance & Development
45. Helliwell JF (2011) How can subjective well-being be improved. New directions for intelligent government in Canada, pp 283–304
46. World Bank Group (2007) City indicators: now to Nanjing. Policy, Research working paper; no. WPS 4114
47. Hou F (2014) Life Satisfaction and Income in Canadian Urban Neighbourhoods. Statistics Canada
48. Hoornweg D (2019) Annual report of the Chief Safety and Risk Officer, Technical Standards and Safety Authority, Province of Ontario
49. Department of Finance (2021) Toward a quality of life strategy for Canada
50. Conway E (2015) The summit. Simon and Schuster
51. Our World in Data (2020) Energy production and consumption
52. Our World in Data (2023) GDP per capita 1820 to 2022; from Bolt and van Zanden—Maddison project database
53. World Bank (2024) US Bureau of Economic Analysis
54. Rubin J (2024) A map of the new normal: how inflation, war, and sanctions will change your world forever
55. Brown G (2024) Finance & development. International Monetary Fund
56. Deese B (2024) The case for a clean energy Marshall plan. Foreign Affairs
57. Friedman G (2010) The next 100 years: a forecast for the 21st century. Anchor
58. Altman D (2011) Outrageous fortunes: the twelve surprising trends that will reshape the global economy. St. Martin's Press
59. Marshall T (2021) The power of geography: ten maps that reveal the future of our world. Simon and Schuster, vol 4
60. Vince G (2022) Nomad century: how climate migration will reshape our world. Flatiron Books
61. Bricker D (2020) Next: where to live, what to buy, and who will lead Canada's future
62. Clarke B, Otto F, Stuart-Smith R, Harrington L (2022) Extreme weather impacts of climate change: an attribution perspective. Environ Res Clim 1(1):012001
63. Lenton TM, Xu C, Abrams JF et al (2023) Quantifying the human cost of global warming. Nat Sustain 6:1237–1247. https://doi-org.uproxy.library.dc-uoit.ca/https://doi.org/10.1038/s41893-023-01132-6
64. Barnosky AD, Hadly EA, Bascompte J, Berlow EL, Brown JH, Fortelius M, Smith AB et al (2012) Approaching a state shift in Earth's biosphere. Nature 486(7401):52–58
65. Wei Y, Wu J, Huang J, Liu X, Han D, An L, Huang J et al (2021) Declining oxygen level as an emerging concern to global cities. Environ Sci Technol 55(12):7808–7817
66. Shaffer G, Olsen SM, Pedersen JOP (2009) Long-term ocean oxygen depletion in response to carbon dioxide emissions from fossil fuels. Nat Geosci 2(2):105–109
67. Armstrong McKay DI, Staal A, Abrams JF, Winkelmann R, Sakschewski B, Loriani S, Lenton TM et al (2022) Exceeding 1.5 °C global warming could trigger multiple climate tipping points. Science 377(6611)

References

68. Osman MB, Coats S, Das SB, McConnell JR, Chellman N (2021) North Atlantic jet stream projections in the context of the past 1,250 years. Proc Natl Acad Sci 118(38):e2104105118
69. Stanley A (2024) Finance and development. International Monetary Fund

Open Access This chapter is licensed under the terms of the Creative Commons Attribution-NonCommercial-NoDerivatives 4.0 International License (http://creativecommons.org/licenses/by-nc-nd/4.0/), which permits any noncommercial use, sharing, distribution and reproduction in any medium or format, as long as you give appropriate credit to the original author(s) and the source, provide a link to the Creative Commons license and indicate if you modified the licensed material. You do not have permission under this license to share adapted material derived from this chapter or parts of it.

The images or other third party material in this chapter are included in the chapter's Creative Commons license, unless indicated otherwise in a credit line to the material. If material is not included in the chapter's Creative Commons license and your intended use is not permitted by statutory regulation or exceeds the permitted use, you will need to obtain permission directly from the copyright holder.

Chapter 7
Brave the Future: The Road Ahead

All flourishing is mutual
—Robin Wall Kimmerer

Big things are accomplished together. The collective wins every time. Cities and civilization highlight the propensity, the innate need, of humans to socialize, and to work together. Traveling together toward the common good, however, is rarely an easy road.

In *Civilization and Its Discontent* [1] Sigmund Freud addresses the fundamental paradox of civilization (i.e., cities): civilization emerged to protect us from unhappiness, and yet it is our largest source of unhappiness. Civilization at times must compromise individual happiness to fulfill its primary goal of peaceful coexistence. The common good is predicated on peaceful relationships with one another, which can overshadow individual urges. Freud acknowledged the ill-will within many hearts, and that civilization exists to restrain these impulses.

This ill-will is clearly visible in the vitriol of social media, the recalcitrance to stay within planetary limits, regionalism, tribalism, growing inequality, and the erosion of reciprocity and global cooperation.

Civility is the glue that binds humans together into disparate, productive groups. This is the greatest contribution of cities and an urban form of life that emerged some 10,000 years ago as agriculture enabled a shift away from a nomadic and more precarious existence.

Planetary limits need to be respected within the limits of social tolerance. And like civility, tolerance grows in cities. Some see this as a noble human aspiration. The pragmatism of the city manager, however, sees this as a necessary ingredient for picking up the garbage and keeping the lights on. City managers, however, are afraid of anger. Anger erodes civility. Like in every other country, Canadians say they are civil, until they are not.

The greatest power of cities is also their greatest weakness. Cities are immobile; they cannot get out of harm's way. And yet cities almost never die. Lost cities like Pompei and Machu Picchu are rare. Examples of cities destroyed and then re-built are more common, e.g., Hiroshima, Nagasaki, Atlanta, Dresden, and Stalingrad. Populations will wax and wane, but cities came before countries, and they will outlast most countries. Countries succeed and fail; cities endure. The flourishing of society will be led by cities.

As outlined in Chap. 1, cities scale superlinearly for things like economy, patents, crime, and pollution, and they scale sublinearly for infrastructure costs. A city of two million people will typically generate more than twice the economy of a city of one million people, and it can do this with less than twice the cost for roads, buildings, water, and electricity distribution systems. This is the main reason why more than 1 million people a week are moving to cities and will continue to do so until at least 2060 [2].

In Canada, however, larger cities will only be able to meet their potential with the express support of all Canadians. The influence of Canadian cities stretches across the entire country, and ever more-so, around the world.

Cities, especially larger cities, provide more opportunities for citizens, and when managed well, can do so for less costs and environmental impact. A challenge however is that as cities grow, they need to innovate, and the pace of this innovation needs to increase as they grow [3]. This is at the core of the growth versus no-growth debate. As cities grow, they consume enormous amounts of energy and materials. Many researchers believe there are finite limits on energy and material use, and that if ecosystem impacts grow commensurately, cities and their national economies cannot grow indefinitely (or innovate increasingly faster).

As American economist Kenneth Boulding quipped, "Anyone who believes that exponential growth can go on forever in a finite world is either a madman or an economist."

With the benefit of 100 years of hindsight and the opportunity of peering 100 years into the future, a possible road ahead emerges. The last 100 years were mostly about growth and geopolitical positioning. This will no doubt continue, but the next

100 years will be more about stability (post-peak population) and durable collaboration. The speed of technological and climate (ecosystem) change will force a more dynamic and hopefully, sustainable approach to human migration and shared resources (i.e., collaborative governance and management systems).

Cities will likely exert greater influence on their countries and each other. Country groupings, like NATO, OECD, G77, G20, the Five Eyes, AUKUS, BRICS, EU, and ASEAN, will continue to form and re-form. The Conferences of the Parties (COPs) under the UN framework will manage international agreements in areas like climate change, biodiversity, and eventually compensation for climate impacts and geoengineering [4].

Cities however will be tasked as first responders to manage several key transitions. Critical among these are the energy and materials transition, and adoption of new technology such as artificial intelligence and geoengineering. In all of these, culture and civility, play a paramount role.

Culture drives human evolution more than genetics [5] and cities will need to facilitate this cultural adaptation. Eric Cline in *After 1177 B.C.: The Survival of Civilizations* illustrates how it is a society's adaptive capacity, the ability to thrive under stress that determines a community's future when facing adversity [6]. And the next 100 years will be full of adversity.

Cities (urban areas) will be forced to create a new type of Hanseatic League, demanding of their countries and businesses (both of which are non-scalable) a more dynamic and flexible approach to human migration, compensation, adaptation, and urban service provision.

When social scientists work with physical scientists amazing things can happen. For example, infrastructure is usually impacted and limited by the peaks: A cold winter day and buildings need much more heat, rush hour's crowded roads trying to accommodate people traveling to or from work and school, or a hot summer day when air conditioning demands overwhelm the power grid.

Engineering a community to better balance demand is just as powerful as building yet more urban infrastructure. A good infrastructure engineer is always looking for help to establish the social support needed to build the power plant and waste management facility or reduce congestion such as by applying time-of-use pricing. A globally minded infrastructure engineer searches for efficiencies, equity, and ways for people and cities to work together.

A transportation engineer, for example, knows that a bicycle is the most efficient locomotion device ever developed (for efficiency of energy and materials, in Canada, a bicycle is at least 1100-times more efficient than an automobile in moving people).[1] The world's cities can readily support integrated mobility with transit,

---

[1] See Fig. 16.4, Bermuda's Delicate Balance (1981), adapted from Papanek, J. and J. Hennessey (1977) *How Things Don't Work*. Pantheon. A human on bicycle uses 0.12 cal/g/km while a fish uses 0.41 cal, horse 0.65 cal, human 0.7 cal, automobile 0.8 cal, helicopter 3 cal, hummingbird 4 cal, mouse 60 cal/g/km. As the average bicycle is 11 kg and automobile in Canada is 1860 kg, and adult is 82 kg, travel by bicycle requires 11,160 cal/km while automobile 1,673,600 cal/km (i.e., bicycle more than 150 times more efficient than automobile).

safe streets, integrated ridesharing and deliveries, carsharing, and bicycles (with e-bikes); however, getting people to shift from single-occupant, personal automobiles is difficult. Few cities have achieved this, although Montreal's initial progress is noteworthy.[2]

The shift in Canada's vehicle fleet over the last several decades illustrates how social norms drive many of today's infrastructure challenges. Canada's vehicle fleet is now the least fuel efficient in the world (8.9 L/100 km) and most polluting (>200 g CO_2/km driven) with vehicles the heaviest and largest in the world [7]. This shift to larger more polluting vehicles is compounded by the removal of licensing fees and road tolls in provinces like BC and Ontario, and reductions in fuel taxes (as well as agitation to remove a price on carbon as part of climate mitigation initiatives).

Chester and Allenby [8] suggest a new set of infrastructure-first principles for the Anthropocene: (1) plan for complex conditions and surprises; (2) recouple with agility and flexibility; (3) govern for exploration and instability; (4) build greater consensus as control decentralizes; (5) restructure to engage with more permeable boundaries; and (6) cyberthreat planning is now mission critical. These principles should guide infrastructure planning recognizing the changing nature and increasingly obsolete boundaries that have defined engineered systems in the modern era.

Over the last fifty years, driven by growth in cities, society shifted from matter is abundant and energy is scarce to energy is abundant and matter is more limited. For example, average copper ore grades globally have dropped from four percent in 1900 to around one percent in 2010 [9].

Eliel Saarinen wrote, "Always design a thing by considering it in its next larger context—a chair in a room, a room in a house, a house in an environment, an environment in a city plan." [10] When designing and managing cities, especially in Canada, designs often reflected the province that the city is in, and maybe the country. Montreal, Toronto, Vancouver, and increasingly Calgary and Edmonton, are mostly influenced by their overseeing province.

The next larger context for cities is shifting from province and country to a global perspective. Climate, supply chains, migrants, peace and order, these are mostly global issues. Countries have not comported themselves well: negotiating limits on greenhouse gas emissions, for example. Cities will continue to strengthen their voice in these areas.

Leadership emerges through good government *and* good management. Urbanization and the cities that arise, force pragmatism, and blur borders. Especially the borders between municipalities that are part of a common urban agglomeration. Cities require sound public policies with day-to-day implementation as they outgrow these borders.

---

[2] London, England, and Stockholm provide compelling evidence for the beneficial impacts of congestion pricing. Copenhagen and Amsterdam highlight the impact of integrated cycling. Many large cities in Asia, such as Thailand, Jakarta, Mumbai, and Hanoi, are transforming urban mobility through e-scooters (with approximately 100 cc engines).

7.1 The Tenacity of Cities

The tenacity of cities is even more impressive considering the inherent fragility associated with urban services. Cities need enormous amounts of imported energy, materials, and food. So much, that large-scale storage is often impractical. Most of the challenges associated with the current energy transition is energy storage. Fossil fuels are particularly valuable as an energy source because they can be easily transported and stored to meet fluctuating demands. However, a city like Toronto or Montreal typically only has two-to-three days of energy (gasoline and diesel) and food available [11].

The high reliability of electricity supply breeds complacency, even though every adult Canadian likely recalls a time that they muddled through a power outage, sometimes even for a few days. Ice storms, regional blackouts, geomagnetic disturbances, wildfires, floods, overtaxed transformers and transmission lines, terrorists and malcontents, all threaten the reliability of electricity supply. Yet overall electricity demand is set to more than double in Canadian cities[3] and supply systems will need to cope with a much more challenging climate. Urban systems, housing, and data transmission are being developed and built with the belief that this extremely reliable electricity grid will continue.

In January 2022, as temperatures plummeted to -30 °C, Hydro Quebec experienced its peak electricity demand of 40,300 MW.[4] Almost half of that supplied Montreal.

Water supply and food is another vulnerability of cities. A large city like one the size of Toronto, Vancouver, or Montreal needs about 1 million m^3 of potable water treated and delivered every day (~5 million bathtubs full). And a city with a population of some four million people, that typically has an additional 400,000 people passing through as visitors, students and temporary workers, consumes about 1500 tons of food per day, and generates 1-to-5 million m^3 of wastewater (depending on precipitation), and 5000 tons of solid waste, or about 400 truckloads of garbage every day [12].

There is much harm coming in the next 100 years. Cities are more reliant on technology like global positioning systems. A few lines of defective computer code can disable systems. Air travel around the world was disrupted with more than 150,000 flights canceled in July 2024, as common software updates failed to load [13]. Similarly, on July 8, 2022, more than 12 million Rogers Telecommunications customers across Canada were denied cell phone and Internet service as the firm experienced

---

[3] For example, for example Ontario's IESO suggests annual consumption rising from 151 TWh in 2025 to 263 TWh in 2050, and Hydro Quebec predicts a more than doubling from 60 TWh in 2035 to between 150 and 200 TWh in 2050 https://www.hydroquebec.com/data/a-propos/pdf/action-plan-2035.pdf (accessed 12–9-2024).

[4] In January 2022 Ontario's peak electricity demand was 21,349 MW, about half of Quebec's demand, even though Ontario has 50 percent more population than QC. However, Ontario cities mostly rely on natural gas for space heating. As net-zero carbon objectives are pursued, space heating will demand more electricity (e.g., heat pumps), and Ontario's electricity peaks are expected to increase.

a major malfunction, lasting more than 24 h. In some communities 911 emergency services were unavailable. The system fault was later attributed to human error [14].

Cities (urban areas) need to enhance resilience. Like how the Government of Canada urges that residents should be prepared to take care of themselves and their family for a minimum of 72 h [15] cities and their urban services such as electricity, water, and telecommunications, need to be sufficiently resilient to act on their own for several days. In large-scale failures, systems engineers try to ensure that components within the broader network can operate independently when necessary. Emergency preparedness planning for urban services is undergoing major changes, as events like COVID-19 highlight the fragility of global supply chains, and the fractious nature between Canada's provinces and federal government agencies and uncertain political leadership [16].

7.2 Pebbles in the Pond

Ottawa, through the National Capital Commission (NCC), first established car-free days along the Capital's scenic parkways in 1970 [17]. The idea was replicated and expanded by the City of Bogota in 1974, and two years later Mayor Luis Pietro Ocampo institutionalized the concept through the permanent establishment of *Ciclovia*. Bogota's main streets are now closed to motor vehicles every Sunday and holiday from 7 am to 2 pm. In the last 50 years, *Ciclovia* was replicated in more than 200 cities across 40 countries, including Winnipeg, which was the first Canadian city to hold a *Ciclovia* in 2009 [18].

The first Business Improvement Area (BIA) started in Toronto's Bloor West Village in 1970. Business owners were concerned about declining retail sales and neighborhood vitality. Under authorizing provincial legislation and city administration, each property owner was levied a fee to help support local improvements. All funds were pooled and administered by the BIA. In the last 50 years, the concept was widely replicated, including most Canadian cities, 84 BIAs (with 45,000 members) across Toronto [19], and more than 60,000 BIAs worldwide [20].

The first curbside Blue Box collection program was launched in 1983 in Kitchener, Ontario. Nyle Ludolph, a strong supporter of recycling, and a senior employee with the waste management firm Laidlaw, proposed to expand the pilot program city-wide as part of a successful waste collection bid. Residents embraced the program that was quickly replicated across Ontario (more than 95% of residents eventually included) and Canada. The program served as a blueprint for recycling programs in more than 150 countries [21].

What is presented as leadership in cities is often 'just' good management. Useful initiatives grow from the 'grass roots' and are quickly replicated. As complex systems, cities present a broad canvas for political leaders, yet a good mayor is often as much a manager as a leader. Cities lead by doing. Cities, as urban areas, exhibit human attributes and are quick to copy each other. Cities derive their agency from the community and compete with other cities, as well as with provincial and national governments. They are also best able to cooperate, a skill that will likely be sorely tested.

7.3 Competition and Cooperation

> *There is unrest in the forest*
> *Trouble with the trees*
> *For the maples want more sunlight*
> *And the oaks ignore their pleas*
> —The Trees, RUSH
>
> This Mother Tree was the central hub that the saplings and seedlings nested around, with threads of different fungal species, of different colors and weights, linking them, layer upon layer, in a strong, complex web.
> —Suzanne Simard
>
> Two tigers cannot share the same mountain.
> —Chinese Proverb

Two views of the forest are possible: Competition, 'red in tooth and claw'; trees stretching to the light, overshadowing those below; and the other, cooperation, connections, like mycelial networks, benefiting all within an ever-changing system whose parts respond to each other and to the environment.

Two tigers may not be able to share the same mountain, yet 197 countries need to share the same planet. How countries do that will be decided over the next 50- to-100 years. Cities and the civilizations that arise from human agglomerations will likely provide the template for how countries prosper together.

Sitting through yet another UN-convened Conference of the Parties (COP), be it for negotiation of a climate agreement (now at the 30th annual meeting), biodiversity (COP16), or desertification (COP16), one is struck by the territorial nature, and zero-sum mindset that pervades negotiations. There are thousands of NGO representatives, corporate lobbyists, and people keen to be part of the circus-like atmosphere. About 84,000 people attended COP28 in Dubai [22].

The main objective of the COP process is to reach a consensus agreement among designated representatives of the 197 'parties.' These parties behave very much like tigers on the mountain. Occasionally country-factions emerge, trying to progress an agreement. For example, the G77 country group of 77 lower-income countries founded in 1964 to provide a collective voice (the G77 now includes 134 member countries).

One of the more influential meetings of the (national) parties last century was the 1944 Bretton Woods Summit that brought together representatives of 44 countries (with about 730 delegates). Key negotiations were between the USA and the UK. The agreement supported much of the prosperity that followed for 60 years; however, the seeds of discontent were sown with many countries resenting the preeminence of US interests, especially after the USA fully abandoned the gold standard in 1971.

With 197 countries in the world, unanimity on almost any issue is unlikely. This is the largest challenge in the current UN-supported Conference of the Parties approach. One country, one vote also seems problematic if a majority is required. And the original genesis of the World Bank and IMF with one dollar-one vote is also difficult as the 2016 establishment of the Asian Infrastructure Investment Bank attests (the AIIB was largely instigated by China, which believed it had too limited a voice at the World Bank).

Perhaps it is easier to suggest sharing of the planet by the 1000 cities expected to reach over one million population, this century. Those cities pay the lion's share of global military operations and buy most of the world's manufactured items and harvested crops. However, if cities are going to assume a greater global voice, they will first need to establish greater harmony with their more rural hinterlands.

The *mouvement des gilets jaunes* (yellow vests protest) originated in rural France, but paralyzed Paris. So too Canada's 2022 'Freedom Convoy' protest that originated in the prairies but closed Ottawa for several weeks.

Recent election returns for Canada and the USA show red, or blue, dots where the cities voted differently from more rural areas (the 2024 US election had a 63%–35% Democrat/GOP split favoring urban areas, and a 36–62% Democrat/GOP split in urban/rural areas; AP Votecast). Brexit and the rise of Donald Trump (twice) were rooted largely in rural discontent. Rob [23] and Doug Ford based in Toronto [24] suggested similar disdain for 'downtown elites' (see Chap. 3). Montreal's voting practices are often at odds with the rest of the province. Alberta's Bill 18, the Provincial Priorities Act, would require municipalities, including Edmonton and Calgary, to obtain provincial approval before entering into agreements with the federal government.

Autocrats and strongmen often fear, or are disdainful of, the city [25]. The world's 1000 million-plus cities will need to build direct, durable and respectful partnerships with their rural hinterlands. This can no longer be left to states, provinces and national governments.

One could argue that Canada became post-urban between 1950 and 1990 when urban tax revenues helped maintain rural communities. The national and provincial governments supported rural infrastructure as a means of ensuring territorial integrity and for political considerations. Electricity, gas pipelines (where practicable), waste and recycling collection, roads, and student busing were provided to most of the rural areas, despite costs typically being significantly higher.

Like Canada's equalization payments between provinces and regions, a general goal of cost parity is applied between urban and rural households. Despite highly subsidized infrastructure, geography prevailed, and rural household costs were still higher. Rural households often had to supply their own water and wastewater systems, home heating, had fewer shopping alternatives (with greater shipping distances) leading to higher prices, and traveled greater distances. Governments compensated somewhat with reduced vehicle license fees, electricity rates, and localized tax reductions.

The 100-year history of urban–rural and regional tensions in Canada may serve as a model for the rest of the world. There will always be regions less affluent, less privileged, and less influential in governance issues. There will always be resentment as one area believes it was discriminated against or believes its culture, language, religion is under threat.

Canadian cities are large, messy agglomerations of local governments, often at odds with their neighbors and their province and federal government. As these cities work toward collaboration, the lessons and learned behavior can scale to the entire world. Cities are nodes within global networks, connected across vast distances. Harm one city and impacts cascade globally. Not just energy and materials but also social networks. Conversely, one city can also beneficially cascade collaboration and sustainability outward.

> Civilization is like a thin layer of ice upon a deep ocean of chaos and darkness.
> —Werner Herzog

Every year in early winter, Canada offers a lake-side seat to system change. As water temperatures drop, ice forms, first gradually, then suddenly [26]. In spring, the process is reversed. Often, with a loud break, the ice is out in just a couple days. A lake changes back-and-forth from ice covered to open water, slowly at first, then all at once.

As outlined in Chap. 2, Dana Meadows suggested the most powerful way to intervene in a system is to change the narrative. Like a lake, open water one day and ice covered the next, the city narrative can change quickly, with inexorable force. Two changes to the city narrative may well start in Canada and may fundamentally change cities, the country, and the world. The two narratives, infused with the wisdom of Canada's First Nations, are *scarcity to sufficiency,* for energy and material flows, and linking migration with sustainability.

7.4 The Roots of Immigration—The Need for Querencia

The first time I came across *querencia* was reading *The Rediscovery of North America* [27] where Barry Lopez captured in one word how people might connect to the land. *Querencia* is from the Spanish verb *querer,* to want, like, love, and reflects fondness, haunt of an animal, favorite spot [28]. Querencia is the place where a person's roots are deeply anchored, where they are most at home.

Ernest Hemmingway in *Death in the Afternoon* wrote "querencia is a place the bull naturally wants to go to in the ring, a preferred locality. It is a place which develops in the course of the fight where the bull makes his home. It does not usually show at once, but develops in his brain as the fight goes on. In this place he feels that he has his back against the wall and in his querencia he is inestimably more dangerous and almost impossible to kill" [29].

Lopez believed the word *querencia* was compromised when used to describe a place the bull goes to in the ring:

> I would like to take this word querencia beyond its ordinary meaning and suggest that it applies to our challenge in the modern world, that our search for querencia is both a response to threat and a desire to find out who we are. And the discovery of a querencia, I believe, hinges on the perfection of a sense of place [30].

In *A Sand County Almanac* [31], Aldo Leopold argued for a similar "land ethic" relationship between people and the land they inhabit, based on respect. "We abuse land because we regard it as a commodity belonging to us," Leopold wrote. "When we see land as a community to which we belong, we may begin to use it with love and respect."

In *The Rediscovery of North America,* Lopez examines the attitudes of 1492, and today, that informed the settlement of North America. "The assumption of an imperial right conferred by God, sanctioned by the state, and enforced by the militia, the assumption that one is due wealth in North America." Lopez goes on to offer hope: "This violent corruption needn't define us. We can take the measure of the horror and assert that we will not be bound by it." We can "rediscover" North America, and the rest of the world—not as a source of income but as a home, a place in which we are to find civility, our strength and character [32], our *querencia*.

My mother and father immigrated to Canada from the Netherlands in 1953, newly married, and hopeful. They grew up during the Depression and made it through WWII, where my dad was held prisoner and needed months of convalescence after the war ended. With my older sister just a year old, the three sailed to Canada and disembarked at Pier 21 in Halifax.

When my parents came to Canada, they largely severed their ties to home and re-rooted themselves in Canada. My mom's main connection to her mother back in Rotterdam was the weekly airmail letter, Christmas parcels to us grandkids, and the occasional, very expensive long-distance phone call. Trips back to Holland were rare.

My parents moved to Trenton, Ontario where they contributed to the Great Acceleration with a family of five children, and eventually 14 grandchildren and 24 great-grandchildren. My parents moved eight times within Trenton, and my dad bought a dozen new cars.

In August 1955, my dad started working with the Town of Trenton and eventually became City Engineer, a position he held until retirement 37 years later. After retiring, he entered politics and served as a local municipal councilor. My mom and dad are buried in Mount Evergreen Cemetery in Trenton, now Quinte West after the 1998 amalgamation, which my father was steadfastly against.

Today people can immigrate to Canada and maintain closer ties back-home. With inexpensive video calls, targeted real-time social media groups, much more affordable air travel, and soon-to-be virtual reality visits, today's immigrants are often well-connected 'back-home'. On balance this is good, but going forward, especially when more than half of Canadians are foreign born (expected around 2041),[5] the ties to Canada, relative to ties around the world, may be more dynamic. Canada has an

---

[5] By 2041 Statistics Canada projects between 49.8 and 54.3% of the population being immigrants and their Canadian-born children. Up from 40% in 2016. Released 2022-09-08.

7.4 The Roots of Immigration—The Need for Querencia

opportunity to be more cosmopolitan, more connected. This is an enormous potential strength; however, strong local roots remain imperative, especially as circumstances are likely to demand greater tenacity and resilience.

International tensions are likely to intensify. Precursors are already evident. Accusations against India for extra-territorial executions in Canada, and China's political influence and pressures on Chinese diaspora are examples. Immigrants may move to Canada, but they typically live in a Canadian city, often in the suburbs or exurbs. Here is where they grow roots. Cities will need to help nurture these roots, while also recognizing that most of Canada's citizens will maintain additional roots elsewhere. Cities provide humanity's first draft of civilization, and the narrative is changing.

Cities will change the narrative of who gets to call Canada home over the next 100 years and how this determination is made. Canada's acceptance of new immigrants and foreign workers, by percent of population, is one of the highest in the world, and recently Canada became the world's most preferred destination for educated immigrants (those who completed at least an undergraduate degree).[6] Several major forces will influence immigration numbers. Wars, droughts, political unrest, and a changing climate will likely continue to drive millions toward more temperate countries like Canada.

Another force affecting immigration is the receptivity to newcomers shown by existing residents. These prejudices and fears, and legitimate concerns for capacity to accommodate newcomers, considering housing availability and services such as health care and transportation, will be balanced against the perennial need for new Canadians to support the economy.

By 2036, immigrants will represent up to 30% of Canada's population, compared with 20.7% in 2011.[7] The role of immigrants and non-permanent residents will intensify in communities like Brampton and Surrey, but also in smaller and more rural communities.

Even though it may be one of the world's last countries to reach peak population, Canada will stop growing in the next 100 years. All governments need to prepare for this eventuality and ideally maximize the potential benefits associated with the transition. Cities experience the ebb and flow of populations and the transition to sustainability most immediately, and most viscerally.

Cities are the haunt of humans. Barry Lopez may have been most at home living with the Inuit in Canada's Arctic, searching for meteorites in Antarctica, or writing from his cabin along the McKenzie River in Oregon. But it was cities where his books were published and sold.

Maurice Strong's computer password was his *querencia*, Lost Lake.[8] One of Canada's stalwarts of sustainability, who waged his campaign from the cities of Rio de Janeiro, Nairobi, Toronto, Stockholm, New Delhi, Beijing, and Ottawa, felt most at home in his log cabin on a small lake by the edge of the Canadian Shield.

---

[6] The Economist magazine. April 15, 2024. Talent is scarce. Yet many countries spurn it.

[7] Immigration, Refugees and Citizenship Canada (2023) Canada welcomes historic number of newcomers in 2022.

[8] Personal communication.

Although the analogy is compromised and ugly to most, the bull returning to his *querencia* in the hope of survival is much like cities today. Cities are wounded, but they are also the cause of egregious harm. The challenge for cities, and most of us, is that the fight comes from within just as powerfully as from outside.

The strength, and vulnerability, of cities is that they are anchored in place. Cities will outlive their countries, but they need to live through their citizens. Citizens are the ones who will take a stand and protect what they love, their families, their city, and their planet.

If asked, my father, a city engineer for 40 years, would likely have given his *querencia* as Bon Echo, Algonquin, and Lake Superior Provincial Parks. The place where the wolves can be.

7.5 From Scarcity to Sufficiency

As the Great Acceleration picked up speed in the 1950s and 1960s, consuming enormous amounts of energy and materials and generating growing pollution, many grew wary of the impacts and likely trajectory [33]. In 1972, the seminal *Limits to Growth* provided a warning: The earth's interlocking resources—the global system of nature in which we all live—probably cannot support present rates of economic and population growth much beyond the year 2100 [34].

Debate and obfuscation followed. For example, fossil fuel companies promoted alternative narratives and withheld corroborating climate modeling [35]. International negotiations focused on the pace of GHG emission reductions, which of course were intractable as the 197 participating countries have widely different objectives and abilities (and consensus is necessary under UNFCCC protocols).

Limits to Growth and related research on planetary boundaries [36] also suggest that as difficult as an agreement will be on reducing GHG emissions, it will be insufficient as climate is only one of several interrelated planetary ecosystem limits. The planet's capacity to provide energy and material resources, while absorbing the impacts of ecosystem degradation and ameliorating waste and pollution is the overall combined limit.

The resource development and scarcity mindset, so typical in Canada over the last 100 years, is giving way to recognition that the planet's provisioning earth systems, e.g., climate and biodiversity, are increasingly unstable. Triggering of large planetary-scale tipping points is likely. Current economic systems and international frameworks are not well suited to adapt to these potentially rapid and massive changes.

Glimpses of the challenge are evident. Climate change is already rendering some regions unlivable and assets uninsurable, driving up home insurance rates by 73% over the past decade [37]. Many Canadian cities have shifted fire protection and emergency planning from an 'inside out' approach (activities within the city) to outside threats from encroaching danger, e.g., wildland fires and overland flooding. Air pollution remains one of the world's biggest killers, killing at least nine million people every year [38]. Humans and our domesticated animals now have a mass of about 1020 Mt, while remaining wild terrestrial and marine mammals make up just 60 Mt (most of that being whales) [39]. The Global Living Planet Index outlines a 69% decline in wildlife populations around the world between 1970 and 2018 [40]. More than a third of the world's forests have been cut down, mostly to grow food, of which more than half is wasted.

As the Great Acceleration shifts to deceleration and eventual stasis (aka the great simplification, metacrises, degrowth, and decoupling) a stabilizing, and eventually declining global population will provide additional impetus to efforts to enhance collaboration over competition.

> Growth does not come from using more and finite resources, but from discovering more and more productive ways of using those finite resources.
> —Daniel Susskind.

7.6 Fracking Our Cities

Geologists know that productivity of a well rests on two things: porosity and permeability. Porosity is the given attribute of geology, the amount of pore space, grain size, composition of the rock, and what fills the pore spaces, e.g., oil, gas, or water. This cannot be changed.

Permeability, how material moves through the ground, however, can be changed. Fracturing the rock, or "fracking", by subjecting it to high pressures, introduces new conduits for material to move. Productivity of the well can increase markedly. The USA changed the geopolitics of oil and gas through fracking, especially with 'tight' shale formations thought to be exhausted to further oil and gas production. Through fracking, the USA is now the world's largest oil and gas producer.

Fracking for increased productivity requires two things. First, the traditional application of hydraulic pressure. This goes back to the late 1800s. The second, more recent, innovative aspect, is horizontal drilling. This began in the early 2000s. Hydraulic fracking and horizontal drilling combined increased the productivity of some wells by more than 20-times.

An argument might be made that metaphorically, Canadian cities need to be fracked. The porosity of Canadian cities is relatively high. Canadian cities and the people and businesses in them are well served by the rule of law, cultural diversity, an entrepreneurial spirit, enforceable contracts, and good basic services. Governance is generally good, with relatively stable institutions, currency, and climate.

The permeability of Canadian cities however is often limited. Drags on productivity include provincial trade barriers, agricultural marketing boards, provincial squabbles, government monopolies such as rail and liquor distribution, inadequate data collection, consistency, and sharing. Oligopolies often exist in key sectors such as grocers, air travel, and telecom.

Parochial and poorly connected local governments and utilities can also limit productivity. Greater Toronto, for example, has more than 34 transit agencies, 17 electricity distributors, 25 school boards, 8 health networks, 25 publicly funded colleges and universities with more than 40 campuses, along with 21 upper- and separate-tier municipalities, and 89 lower-tier municipalities. These agencies often compete more aggressively with each other rather than with their global competitors.

Most of Canada's cities are divided and subdivided, with services rarely consistent with their urban agglomerations and borders. Montreal is made up of 82 local municipalities. Vancouver has 23 local authorities. Permeability across these government and agency borders can be low, limiting coordination, and reducing productivity.

Breaking these barriers and enhancing permeability (productivity) likely does not require yet another level of government, or a new agency. Enhancing collaboration among existing levels of government and their delegated agencies is the key. Fracturing silos and ossified hierarchies could significantly enhance productivity. Local governments and their agencies are best positioned to identify and take advantage of these potential opportunities. Governments need to communicate to all Canadians that the country, and everyone in it, benefits when cities, especially larger cities, increase their productivity.

The national and possibly provincial governments could share a portion of the goods and services tax (GST, along with the provincial sales tax, PST, combined in some provinces as the harmonized sales tax, HST) with the five largest urban areas provided that their economies are growing (as measured by GDP in the CMA) at a rate higher than the national average, while also reducing the intensity of GHG emissions, solid waste, homelessness, and poverty.[9]

Continuing the analogy of increasing productivity by fracking an oil and gas well (or water well) using high pressure along with horizontal drilling, Canada's cities are blessed with high potential ground conditions. Cities are typically already fracked through the pressures exerted by existing governments, agencies, and managers. The catalytic value that came about through horizontal drilling in petroleum wells can be replicated for cities by requiring contiguous communities, and their agencies and utilities, to collaborate. Enhanced collaboration can be measured directly through increased productivity. Those communities collaborating more, would be recognized, and rewarded more.

Canada's five largest regions are on track to generate more than 75% of the national economy (up from today's 60%). Enhancing productivity in these five urban areas will benefit the entire country, and hopefully the whole world. Productivity, however, needs to be measured beyond GDP alone. Economic growth cannot come at the expense of increased ecosystem damage or growing inequity (equality, well-being, and resource use—circularity and emissions—need to also be monitored by the delegated communities).

Countries and businesses invest in research and development, often without city participation, even though most of the new technologies and processes they might discover are destined for urban areas. However, the demise of Toronto's Quayside experiment with Alphabet (Google's parent company) highlights the challenges of

---

[9] Arguably St. John's, Halifax, Moncton-Saint John, Quebec City, Winnipeg, Regina-Saskatoon, Kelowna, Victoria, and other CMAs could be added. Funds should be allocated annually based on relative increase of GDP, by CMA, published annually by Statistics Canada.

developing R&D in real-time with specific communities (see Box 3.1). Research and application in cities must focus on people, and needs to be inclusive, and ideally community centric. The work of Google was not.

The next phase of global development needs to grow the economy in urban areas, by enhancing security, services, and economy, while using fewer materials and less energy. Developing robust partnerships with cities is critical (local governments but also other agencies of cities), however cities are not well-versed at partnering with the private sector. They also have difficulty convincing residents of the need to govern for the commons and benefits of longer-term solutions. A credible, ongoing suite of understandable metrics, e.g., local SDGs, might be able to help make the case for more comprehensive and long-term partnerships.

7.7 Tear Down These Walls

The Berlin Wall, China's Internet firewall, and the wall along the southern US border; walls can often require significant resources to establish and maintain. Like the Great Wall of China and Hadrian's Wall, circumstances change, and the utility of a long-held wall may also change.

Canada's cities can be viewed through the walls that act upon them. The walls emerge along numerous borders, reflecting local aspirations and insecurities, business zones, and service delivery areas (political and contractual).

The two most apparent walls or borders are national and provincial (territorial). The national border is the wall that limits migrants, defines the currency, and through the national government, projects power globally. Canada's size of just over 40 million people and less than 1.5% share of the global economy (and declining) require a rules-based, multilateral approach to geopolitics. Geographically, with three relatively difficult to access coasts and the world's most powerful country to the south (and along the Alaska border), Canada has not yet been required to actively defend its sovereign territory.

The main international wall, the US-Canada border, is one of the world's most permeable international borders. This is readily verified standing at the Windsor-Detroit border while up to 40,000 vehicles cross every day (with about $350 million in goods). This relatively permeable wall is also evident at airports in Vancouver, Edmonton, Calgary, Toronto, Ottawa, Montreal, and Halifax as passengers can pre-clear US customs. The US-Canada border may undergo tightening, e.g., with the second Trump administration; however, long term, this international border is likely to remain one of the more permeable.[10]

There are many national efforts to differentiate commerce within Canada. For example, agricultural marketing boards limit foreign products, thereby protecting farmers and companies in Canada while increasing prices. Books (including e-books) typically have higher Canadian prices (above potential currency exchange impacts),

---

[10] Based on common culture, language and business integration.

and in some cases, products are unavailable. The recent loss of Kleenex brand in Canada is illustrative. Kimberly-Clark, the parent company, cited "unique complexities" for discontinuing the brand in Canada [41]. Kleenex joined a long list of products that left Canada citing supply constraints, profitability, and local market challenges.[11]

The permeability of the US-Canada border is countered with relatively impermeable, thick provincial borders. Several commodities and professions move more easily across the national border than between provinces. For example, far more Canadian nurses move across the federal border every day than between provinces. Each province has its own public safety regulator, e.g., for elevators and fuels, professional licensing bodies, income tax structure. The recent fracturing of Canada's professional accounting association into provincial groups illustrates this well. The costs of this increased friction are considerable (as much as 3.5% of GDP).[12]

Oshawa and Milton vie for the status of being Canada's fastest-growing city, each growing about 20% in the last five years; both are about 50 km from downtown Toronto, but both are integral parts of Toronto Region. Distant from any large population center, Dryden and Fort Frances, on the other hand, have seen populations decline by almost five percent over the same period.

This system of cities is driven by global forces, even more powerful than national forces. Cities, combining population and wealth, drive the zeitgeist more than countries do. Cities are where new fashions originate, where patents are developed and registered, where IP is created and protected. Musicians play in cities. Major conferences are negotiated in cities. Universities, libraries, and museums are all housed mostly in cities.

In Ontario in 1966, 2.6 million people lived in rural communities, making up 37% of the province's population of seven million. By 2021, Ontario's total population doubled to 14.2 million, while the rural population remained relatively flat in absolute terms at 2.5 million people (shrinking to 17% of the overall population) [42].

There is an irony that most of the international airports in Canada are not located in their namesakes. Toronto's airport is in Mississauga. Vancouver's is in Richmond and neither of Montreal's two airports are in Montreal. If urban tribalism must be exercised by residents, their city-tribe should be the name of the city that flashes on the screen when they are checking their flight status home, when standing in the airport in Europe or Asia. Airports are not typically located in Europe or Asia, but rather in Paris, London, Singapore, Beijing, Jakarta, Miami, Chicago, Los Angeles, and so on.

Canada's global influence is waning. The relative ranking of Canada's main cities in the global list of large cities is dropping quickly. Toronto, the country's largest city, reached a high as the world's 40th largest city (by population) around 2000. By

---

[11] For a longer discussion see: Mandel-Campbell, A. (2007) Why Mexicans Don't Drink Molson: Rescuing Canadian Business from the Suds of Global Obscurity.

[12] The border effects persist, with an implied *ad valorem* tariff equivalent of 6.9%. From Statistics Canada 2017 (Robby K. Bemrose, W. Mark Brown and Jesse Tweedle, Economic Analysis Division). National GDP could rise by $80 billion, or 3.8%, and average wages would rise by 5.5%, through reduced provincial tariffs.

2100, despite growing by another 10 to 15 million, Toronto will barely be in the top 100 cities. Vancouver and Montreal will both have dropped below 200.

Walls are designed, with varying degrees of success, to keep something on one side of the wall. But like water behind a dam, entropy is always at work, and energy and effort are required to maintain the wall. Walls and boundaries also impact the mindsets of people on either side of the wall. For example, if a border is effective at keeping people apart, there is less likelihood that in the event of an emergency on either side of the wall, people from the other side will readily come to the assistance of those in need. Fortunately, across North America utility workers often rush to areas impacted by disaster, knowing that their turn may arise in the future. Greater permeability across walls almost always yields greater productivity.

The Canadian soul is often thought to exist among the beauty of the Rocky Mountains, along the Pacific Coast, perhaps within the lakes and woods sprinkled atop the Canadian Shield, or in the vast skies and land of the north. But the soul of Canadians must also surely be anchored to the cities. Similar to how more than half of Canada's indigenous population now lives in cities, yet there is a visceral, unalienable bond to the land, the soul of Canada is manifest in the country's cities but is also rooted to areas across the planet.

7.8 From Metropolitan to Sustainability Mindset

Iveson and Eidelman argue convincingly that we need to shift our thinking from a municipal to a metropolitan mindset [43]. "The metropolitan mindset means moving past the traditional, zero-sum logic that pervades local politics, where city leaders compete with their neighbors for scarce resources, to one that inspires, enables, and sustains collective problem solving across municipal borders." This zero-sum, competition for resources is also prevalent in international relations.

The challenge of Canada's cities, and Canada overall, is not unique. Civilization is a fine balance between competition and cooperation. Larger cities, especially, need a metro mindset as they outgrow their municipal borders, and within the next few decades as they shrink. Also, as cities grow, they often threaten their provincial/state and national governments. A metro mindset recognizes that human nature changes slowly, and everyone needs convincing now and then that we really are all in this together.

The metro mindset appreciates the fallibility of national governments and the need for a global mindset as well as the pressures faced by local governments and their municipal mindset. The metro mindset needs to be anchored in the community, able to think globally and locally. The metro mindset is predicated on collaboration and a fulsome understanding of the dynamic nature and permeability of borders. Workers, shoppers, residents, and students, all need to cross borders, just as carbon dioxide, air pollution, fish, pathogens, and birds cross those same borders.

So much of human history is about competing for scarce resources. Cities are the main cause of this conflict as they drive demand for resources. Energy, minerals, food, and people: the demand arises in cities because this is where the money is to pay for these resources, and from where military power is projected (and again, paid for).

However, as mentioned in Chap. 5, peak oil, or peak lithium (or whatever material) will not be reached because the world runs out of the ability to supply this material. Rather, peak oil and the other peaks will be determined by demand, which arises from people and businesses in cities. Peak coal will likely be reached this year or next [44], peak oil and gas will be reached before 2050, and perhaps the most important peak of all, peak population, will likely be reached around 2085.

The scarcest resource today is the planet's ability to handle the additional byproducts of energy and materials consumption, and society's ability to exist peacefully as genuine and perceived inequities grow.

Salman Rushdie proposes that there are "hinge moments" in history; when everything is in flux and the future is up for grabs [45]. Through a 200-year perspective, specific moments emerge as inflection points for Canada's cities. One such example is the end of WWI, and the drawing of lines on maps. Largely artificial and often driven by colonial powers hoping to maintain power, and wealth, these new countries gave rise to a century of geopolitical turbulence.

Another hinge moment, as WWII ended, was the League of Nations being formally dissolved in 1946, establishing the United Nations in 1945, and the 1944 Bretton Woods Summit that dictated much of the world's financial architecture and creation of the International Monetary Fund and World Bank.

Other hinge moments include the 1973 oil embargo (bringing about a 300% increase in global oil prices), widespread availability of personal computers (1970s), ubiquity of the Internet and the end of the Cold War (1990s), and terrorist attacks of September 11, 2001.

Most people believe the world is now passing through several hinge moments. Climate change and possible triggering of earth system tipping points is a likely candidate. The combined economies of BRICS countries surpassing those of the G7 in the mid-2040s [46] is another, along with the advent of artificial intelligence, and reaching peak global population.

A hopeful hinge moment may be the codification of human nature, recognizing how humans prevailed over other species because of their ability to cooperate, civilize through cities, and share narratives that facilitate scaling and efficiencies of networks and platforms. As Sigmund Freud suggested, civilization is perpetually threatened with disintegration [1]. This may also be manifest through the inexorable laws of entropy or the baser instincts of humans, as suggested by Moloch (the god of unhealthy competition) [47]. Cities may emerge as the means to hold at bay this disintegration and degradation, providing a system of systems, or civilization of civilizations (as a balance to country and regional competition). The tragedy of the (global) commons may best be addressed through the city as commons. This is the foundation of the metropolitan mindset. The sustainability mindset extends this across the globe, recognizing that environmental, economic, or social degradation anywhere is eventually felt everywhere.

7.9 The Twenty-Ninth Day

Lester Brown, environmental analyst and founder of the Worldwatch Institute, outlines in *The Twenty-Ninth Day* how schoolchildren in France were taught the nature of exponential growth with a riddle. A lily pond contains a single leaf. Each day the number of leaves doubles—two leaves the second day, four the third, eight the fourth, and so on. "If the pond is full on the thirtieth day," the riddle asks, "at what point is it half full?" Answer: "On the twenty-ninth day" [48].

Brown goes on to opine "how the global lily pad in which four billion of us live may already be at least half full" and "within the next generation, it could fill up entirely." The Twenty-Ninth Day was published in 1979, five years after the world's population had reached 4 billion. World population reached 8 billion on November 15, 2022 (date officially recognized by the United Nations). The world has passed the twenty-ninth day, and human population will not again double. Peak population of 10.3 billion is expected in the mid-2080s.

Even though peak global population is 50-to-60 years away, countries are now undergoing massive shifts in populations. These shifts in population are overlain with a changing climate, increasing migration, and enormous economic shifts.

Cities and countries will continue to compete for migrants, as well as emplace barriers to keep other people out; however, as the nexus of planetary systems degradation and peak population emerges, a metropolitan mindset may yet appear. The metropolitan mindset replaces the zero-sum mindset when communities recognize that scarcity, or hardship, in one community, eventually manifests in all communities. The sustainability mindset emerges when communities, with metropolitan mindsets, connect across regions and countries.

7.10 Canada's New Narratives

As outlined in Chap. 2, the most powerful way to intervene in a system is to change the paradigm and create a new narrative. For the last 100 years, Canada's cities adapted to national and provincial narratives. Canada's larger cities, like Montreal, Toronto, Vancouver, and Ottawa, saw common narratives often split across local governments and muddled by broader provision of urban services like transportation, energy and telecommunications.

Canada's cities (communities) will need to lead in the establishment of several new and expanded narratives. Two narratives are critical, migration within sustainable development and a shift to sufficiency.

Regarding migration, two significant trends are emerging. Countries like Canada with high reliance on immigration for continued economic strength may face growing competition to attract migrants as global population declines become evident in the next few decades (e.g., change in rates of growth). Countering this trend is the expected increase in climate migrants. Canada's temperate climate and freshwater resources will likely ensure the country remains a top destination for those abandoning hotter and more water-stressed areas.

Canadians know that the country needs to continue welcoming immigrants. The numbers and methods of selection will be increasingly debated, and refinements are likely, e.g., establishing a more formalized Immigration Levels Plan [49] with long-term projections, providing key cities with an opportunity to comment on methodologies and numbers (in addition to the provinces). However, climate and demographics are changing quickly. Within a decade or two, foreign-born Canadians and their children will be the majority, while foreign interference in elections is increasing.

The World Migration Report 2024 estimates that there are 281 Mn international migrants, 169 Mn migrant workers, 35.4 Mn refugees, and 71.4 Mn internally displaced persons [50]. These numbers are rising. Unlike Europe and the USA, Canada's geography limits the ability of migrants to enter the country illegally. Immigration levels, although among the highest in the world, have been largely determined by the national government. The Institute for Economics and Peace predicts that 1.2 billion people could be displaced globally by 2050 due to climate change and natural disasters [51].

Migration, the movement of humanity, is likely to increase and should be seen as a key component of sustainable development (for Canada and globally). Communities of origin, communities of destination(s), and the people migrating (adults and children), all need to benefit from the process to assure sustainability. Canadian communities could lead the shift in mindset. For example, supporting the communities from where doctors and nurses migrate from; ensuring remittance fees are as low as possible; managing cities for dynamic populations and better mixing of new and long-standing residents. Countries have emerged as the keepers of the borders and regulators of migrant numbers. In Canada, with varying degrees of input from the provinces, the federal government determines and manages migration policy. Cities could develop the new narrative where their ability to accommodate new residents is better integrated with similar abilities in other cities. This could be similar to how businesses in one Toronto neighborhood developed an idea (BIAs) that was replicated more than 60,000 times around the world.

The second new narrative, perhaps even more challenging than the first, is to promote and implement a shift from a scarcity to sufficiency mindset. Cities, especially Canadian cities, can lead in managing several intractable political problems. Health care, migration, community resilience through demographic and climate changes, transitioning to sustainability; these are wicked problems (Chap. 1) that require both political leadership (governance) and management excellence. Most of these issues have been entrusted to national and provincial governments, often referred to as 'senior levels' of government. Yet cities are almost always older than their host country. And, if as is feared by an increasing number of earth scientists, we are entering an era of significant variability and possible economic collapse (as the economy is predicated on a well-functioning ecosystem), cities, will be tasked to lead communities through the transition.

7.10 Canada's New Narratives

Managing the sustainability transition will fall mostly to cities. However, the current governance structure of cities in Canada is not fit for purpose. Transportation and housing are two issues that illustrate current limitations in managing the sustainability transition. Managing issues through the sustainability transition is a systems problem, yet urban governance, and increasingly urban management, are addressed through traditional, linear problem solving.

Financial support to public transit (mostly a management issue) is provided by provinces and the federal government with no certainty. Capital projects are often overseen by the province, or federal government, often not well integrated into the existing system. Transit is also addressed in isolation of overall mobility. Powerful tools, like congestion pricing, charging per distance traveled, and provision of 'surface subways', are not included as it is not yet any specific government's mandate. The federal and provincial governments work together to attract electric vehicle manufacturing to Canada (as automobile manufacturing a large portion of the economy), yet even with that enormous financial contribution, community priorities like common recycling of batteries, reducing the mass of vehicles, and role of EVs in integrated mobility are not addressed.

Housing is another example of where the national government decides on the annual intake of migrants (with differing rules of residency and community engagement), the provinces oversee professional accreditation and academic institutions (who often use foreign student numbers to augment budgets), and local governments, often responding to neighborhood pressures, limit new and renovated housing options. Housing in Canada is also a mainstay of wealth creation, and interest rates, which are intended to influence the overall economy, have a disproportionately large impact on new housing starts and affordability.

Giving cities new powers is likely not the answer to intractable problems, nor giving a larger voice to mayors. Cities are even more fractious than provincial and federal governments and the mayors of Canada's larger cities typically represent less than half their urban community. Local governments, and their utilities and agencies rarely work for the good of the whole if it is contrary to their individual priorities. Rather, the shift in narrative is that the community needs to oversee the sustainability transition. Leadership aspects will continue to be delegated; however, a more dynamic form of governance and management is needed.

A priority is defining the city or community. The best option available is Census Metropolitan Area (CMA) as defined by Statistics Canada. Governing and managing a CMA is different than a city (although the local governments of Calgary and Edmonton are largely contiguous with their CMA). In the case of Toronto, at least three CMAs need to be considered in the urban area (CMAs of Toronto, Oshawa, and Hamilton) and two in Vancouver (CMAs Vancouver and Abbotsford-Mission). The Ottawa-Gatineau CMA spans two provinces.

The permeable borders associated with CMAs are replicated globally. Canada's prosperity at the expense of China or Nigeria, in the long run is unsustainable.

Canada's provincial trade barriers highlight this well—as does the fact that considering the clear case to reduce these barriers, they still exist. Managing for sustainability requires a 'sum of the whole' approach. The 'whole' is not as clearly defined as a local, provincial, or federal government border, yet it can still be as important.

Managing the sustainability transition requires a focus on the 'whole', e.g., CMA, while also recognizing that individual parts, e.g., neighborhoods, local governments and utilities, and larger areas, e.g., provinces and regions across the country, must also benefit (and recognize that they are benefiting) in efforts to support the CMA.

With no politician, government employee, institution, or delegated authority representing Canada's CMAs, everyone needs to support them. The change needed to support an idea, not necessarily a specific place. The sustainability mindset replicates this approach across cities in all regions and countries.

7.11 Braving the Futures—Diverging Futures for Cities and Countries?

As outlined in this book, Canada's most influential regions are census metropolitan areas (CMAs). CMAs as spatial areas, reflect the realities of energy and material flows of urban systems, as well as the economic and travel patterns of citizens. Even though most of Canada's 41 CMAs have no direct like-for-like political representation or employees, they are home to most of Canada's future. For some cities (urban areas), like Toronto and Vancouver, they are made up of two or more contiguous CMAs. Toronto Region, also known as the Golden Horseshoe, which will drive much of Canada's future prosperity, is made up of nine CMAs that often bristle and abrade along shifting borders, as well as being subject to suboptimum dictates from provincial and federal governments.[13]

The fact that Canada's future is represented mostly by ephemeral lines on maps may initially be accompanied with consternation. However, connections are growing across urban regions like Vancouver and Cascadia, Calgary-Edmonton, Toronto Region and its role in the world's largest megaregion of the Great Lakes Region, Ottawa-Gatineau, and Montreal with links to the USA, such as along Interstates 81, 87, and 91. The future of these urban regions may well diverge from their respective

---

[13] A few examples include: sale of Hwy 407, disallowing tolls on the DVP and Gardiner, friction between GO and the 34 regional transit agencies, two publicly funded school boards, giving municipalities veto powers over new waste disposal and energy generation facilities, 40-plus year delay on the Pickering Airport decision, arbitrarily removing bike lanes, encouraging post-secondary institutions to compete rather than cooperate, sale of CP and CN railways (with sale of track alignments), lack of coordination on immigration levels, ad hoc dispensation of lands within the Greenbelt, subsidizing the automobile industry and electricity sector, as well as removing licensing costs of vehicles, at the expense of more complete communities. Both the federal and provincial governments also operate through high deficit finance—this debt will mostly (>80%) need to be paid by urban centers.

7.11 Braving the Futures—Diverging Futures for Cities and Countries?

provincial and national governments. The future of nation-states, i.e., beyond 2050, is less certain than urban regions.

The Johari window, first raised in 1955[52] with its known knowns, known unknowns, and unknown unknowns,[14] applies to Canada and its cities.

Many of the known known challenges are understood within ranges. The date and intensity of the peaks still need to be defined. For example, peak human population, which will drive peak waste, peak GHG emissions, and peak energy demand. The coming challenges of climate change are well known; however what the final global temperature increase might be, or what tipping points will be triggered, and when, is not yet known. Knowledge in geoengineering is growing; however, how it might be applied is still unclear. The specifics on how the space industry will develop and be applied is also not yet known, but its growing role in society is clear.

Several important known unknowns for Canada's cities include a measure of Canada's continued willingness to accommodate new immigrants and bring in hybrid visas and residency arrangements. Another is the extent of geopolitical fractures and disruptions to supply chains (both in terms of materials arriving to Canadian cities, and Canada's ability to export materials). Important known unknowns include shifts to technologies and social acceptance and norms. For example, in *Pitfall: The Race to Mine the World's Most Vulnerable Places* [53] a strong case is made for leaving gold in the ground (facilitated through crypto systems for value retention).[15] Changes like this are possible, but only with broad social support.

Cities (urban regions) and countries will likely continue to diverge on how they advocate and respond to broad technical and social issues changes the next 50-to-100 years. How these new technologies and systems are discussed, regulated and managed, is most determined by trust.

David Johnston in *Trust: twenty ways to build a better country* [54] provides a glimpse of how diplomacy and trust operates between specific people and is often anchored in cities. Qatar was lobbying member countries, arguing that the slow visa process, high taxes, and language laws of Montreal warranted moving the International Civil Aviation Authority and International Air Transport Association headquarters from Montreal to Doha. These two organizations were established shortly after WWII, recognizing the vital role Canada played in aviation during the war. Johnston cited Montreal's fifty-plus years of history, and the role McGill University's faculty of law plays in the international air and space law institute. A few days after Johnston's personal intervention with the Sheikha, Qatar ceased its advocacy to relocate the headquarters.

Organizations are losing trust as quickly as other sectors, such as private corporations and governments. Examples include the UNFCCC Conference of the Parties (COP) shifting every year (with up to 80,000 participants) and the Olympics (with widespread corruption identified in the International Olympic Committee).

---

[14] Made famous by a response United States Secretary of Defense Donald Rumsfeld gave to a question at a U.S. Department of Defense (DoD) news briefing on February 12, 2002.

[15] See also, 'The case for leaving gold in the ground'. By Christopher Pollon, Globe and Mail. October 6, 2023.

To enhance trust in key international agencies, the federal and provincial governments of Canada, and Canada's key cities, should encourage more permanent establishment (less international travel) of international organizations and meetings. For example, the following is proposed:

- Canada should encourage a permanent location for the Olympics (probably Athens). So too with the UNFCCC's COP (perhaps Berlin with five-year updates in Nairobi); updating and monitoring sustainable development goals (SDGs) in Rio de Janeiro; select cities to host COP desertification and biodiversity agreements.
- Canada should advocate to host in perpetuity (in Montreal) ICAA and IATA with an expanded scope in space law and management (and geoengineering). Toronto should continue to host the Prospectors and Developers Association of Canada (with large-scale annual conference). A parallel organization to PDAC should be established that monitors and overseas critical materials (a complement to Paris-based International Energy Association). Toronto may also be an excellent candidate city to host an organization tasked with artificial intelligence regulation and shared development.
- Vancouver hosted the Habitat I Conference in 1976 that resulted in the Vancouver Declaration and establishment in 1977 of UN-Habitat (headquartered in Nairobi). Vancouver should host, alternating with Nairobi, UN-Habitat and World Urban Forum conferences. Vancouver should also emerge as the anchor-city for Cascadia and its Asian port.
- Calgary and Edmonton are likely to merge into a more unified linear urban region with Red Deer in between. Calgary should be supported to serve as home to an organization that monitors, and encourages, the low-carbon energy transition (and other impacts). Edmonton should be supported to host an organization that serves as a port to northern Canada.
- Either Winnipeg or Ottawa-Gatineau should be supported to establish an organization to monitor and support climate migrants and international rights in human mobility.

7.12 Out with a Boom?

Potentially, one of the most powerful known unknowns to impact Canada's, and the world's cities in the next 100 years will be how the baby boomer generation bequeath their wealth and wisdom to subsequent generations. USA and Canadian baby boomers (born 1946–1964) are on track to transfer some $90 trillion between generations.[16] Significantly more wealth will be generated in the next 100 years (see

---

[16] New York Times. May 23, 2023. The Greatest Wealth Transfer in History Is Here, With Familiar (Rich) Winners. Accessed 11-13-2024. The $90 trillion value is derived from $84 trillion in US NYTimes article), and an estimated $6 trillion in Canada.

Chap. 2); however the timing and relative scale of this wealth have the potential to shift social norms and infrastructure, especially in cities.

The baby boomers will not only pass on wealth. Their wisdom and collective behavior will also echo through communities for decades after their departure. The 2024 US presidential election is perhaps a harbinger of potential impact. Voters 50-plus favored Trump over Harris, 52–47%. This is a four-percentage point shift from 2020 which was attributed to being sufficient for Donald Trump to win.[17]

Elder movements, like those more traditionally seen in Indigenous societies, may emerge with a focus on philanthropy, and a more inclusive, environmentally supportive society, e.g., The Elders (theelders.org) and Elders Action Network (eldersaction.org). These seniors working together and through their associations might catalyze enhanced collaboration and support cultural and educational initiatives. A precursor for this movement might be the seen in family foundations such as the Masseys, McConnel, Carnegies, and Rockefellers, and institutions like McGill and Cornell University, Massey Hall, and the United Nations headquarters.

How demographics and rapidly shrinking populations impacts countries like China, Japan, Russia, and much of Europe in the next several decades, and how these impacts spill out globally, will also dramatically affect Canadian cities (see Chap. 2). Much of today's friction between China, Russia, and the OECD-aligned countries today reflects increasingly rapid demographic shifts. The inexorable aging of human populations, more acute in some countries, with the speed at which peak population arrives, and the rate of decline after the peak, will be one of the most profound impacts on world cities, especially those in Canada.

7.13 Municipal Pooling

A municipal myth, a finer version of an urban myth, for the City of Toronto maintains that the city, justifiably proud of its municipal water quality, and concerned with single-serving beverage containers, proposed a new business line of retail water sales. The idea was not pursued, in deference to the retail and beverage industry. Municipalities are typically not encouraged to establish businesses that might compete with the private sector.

A few municipalities explore innovative partnerships with the private sector. Innisfil, a sprawling, fast-growing, mostly rural municipality of just under 50,000 people, south of Barrie, Ontario, launched in 2017 a partnership with Uber to provide residents a unique ridesharing option. The subsidized trips were thought to be cheaper than provision of municipal transit, with infrequent, mostly empty buses. The successful pilot concluded in September 2024. Deliberations continue how best to continue the popular, but costly, program.

---

[17] AP Vote Cast as reported by AARP, https://www.aarp.org/politics-society/government-elections/info-2024/election-analysis-older-voters.html (accessed 11–13-2024).

On an average day in an urban residential neighborhood of a large Canadian city, post-COVID-19, some 40-to-50 delivery vehicles pass every house (Toronto averages 23,600 active vehicles daily, providing about 212,000 trips). Like how postal services developed around the world, supporting efficiencies through consolidation, municipalities will need to drive more efficient use of city streets.

Other areas where municipalities may drive increased efficiencies and community well-being include: supplying every household with an electronic tablet (or Internet application) that provides real-time information and emergency communications; consolidating and coordinating neighborhood level interventions (and specific households) to increase resilience (and reduce insurance premiums for participating households); consolidated deliveries of commonly purchased items, particularly for seniors and shut-ins; a broader spectrum of integrated mobility.

Another area where municipalities may need to intervene is enhancing subjective well-being (i.e., happiness). For example, as outlined in the 2025 World Happiness Report, one of the best ways to increase happiness is to eat with someone else. Like the annual Dine Out Vancouver (starting 2002) and Winterlicious/Summerlicious Toronto, restaurants might be supported to provide community tables to encourage collective meals. Targeting these programs to include youth (<30) would be especially useful to increase happiness.

Municipalities and their utilities employ about 2 million Canadians. No other group has greater potential to directly improve the well-being of Canadians. Empowering this group to serve the public to the best of their abilities requires broad public support and a shift in attitudes on what municipal (and utility) employees can contribute to the community. This potential enhancement of civil service also applies to the provincial and federal (and international) level;[18] however shifting attitudes and behaviors is likely best applied from the ground up. Enacting a municipal employee recognition system, to support innovation and excellence, would be an important addition to Canada's communities.

Richard Marceau and Clement Bowman (2012) highlight in *Canada: Winning as a Sustainable Energy Superpower* how Canada was knit together by big, shared (pooled) infrastructure projects. Namely: Rideau Canal, Victoria Bridge, Canadian Pacific Railway, TransCanada Airline and microwave, St. Lawrence Seaway, Canadian satellites, James Bay Hydroelectric, Oil Sands, Nuclear Power, synthetic rubber, and natural gas pipelines. They argue for the next iteration of "big projects" through hydroelectric and nuclear expansion, and development of a 735 kV Pan-Canadian transmission grid [56].

In Canada's provinces and territories, cities drive most of the electricity (and other energy) demand. Partly because of their fractured nature, e.g., Toronto Region's 100+ municipalities and 30+ local distribution companies, cities have relatively little influence on Canada's energy infrastructure. Developing future energy infrastructure requires more active involvement of local governments and First Nations communities. Cities are best positioned to aggressively promote energy conservation (efficiency) and to help negotiate host community support for infrastructure. Canada's

---

[18] For example, see Lewis [55]

provinces will continue to advocate for developing and selling energy, e.g., fossil fuels and electricity; however, cities will help drive greater energy security and equity. Pooling the aspirations and capacities of municipalities is a powerful way to speed up energy infrastructure development.

7.14 Recommendations

As Canada's cities and urban communities face the triple planetary crises (climate change, pollution, biodiversity loss), struggle to shift to a net-zero carbon economy, maintain and build a nurturing society at home and abroad, and continue to contribute to Canada's prosperity, the following recommendations are proposed:

New Approaches to Housing

In the early 1900s, about a third of Canadian household members were unrelated. Over the last century, this dropped to less than a tenth (Chap. 2). By 2030 about a quarter of Canadians will be seniors (this ratio is increasing by about 1.5% per year; about 25% of people over 85 live in a senior's home).

International students and recent immigrants (plus a growing number of climate migrants) continue to exacerbate housing demand. Baby boomers are exhibiting a tenacious desire to stay in their homes as they age, often with larger space availability than the number of residents warrants. Based on 2021 census data, the Canadian Center for Economic Analysis (CANCEA) estimated that there are 5 million empty bedrooms in Ontario alone. Canada would likely be about twice that, or 10 million empty bedrooms.

Over the next 50 years, city populations will ebb and flow markedly. Some cities, like those in Europe, will decline in population. Some cities will also likely emerge as staging centers, serving as nodes for migrants, and transitional residents.

The expected widespread changes in circumstances suggest that local zoning requirements should be modified to help cities to encourage more dynamic living arrangements. Culture and expectations may also need to change, especially as many of the upcoming changes may manifest rapidly. Municipalities should take the lead in introducing web-based searches to optimize potential matches while minimizing safety concerns and possible neighborhood opposition.

Chief Resilience (Sustainability) Officers

Chief Resilience Officers should be established for (at least) the five major urban regions of Canada. The CRO's mandate should include an annual report, plus five-year master plans on systems sustainability. Assessments should focus on two broad areas—infrastructure and critical systems, and social services and community preparedness. CROs should not report to a single municipality or province, but rather to a board made up of representatives from local, provincial, and federal governments, utilities, and public health officers. The office of the CRO should not exceed a few staff (with a modest long-term sustained budget). As a minimum, the

CRO's mandate should cover the broader CMA, and in the case of Toronto, should include all Toronto Region (Golden Horseshoe).

Enhancing resilience requires a systems approach. The resilience of Calgary, for example, is dependent on Airdrie next door, and support from the provincial and federal governments. An important contribution of the CRO should be an annual workshop where representatives of local governments (CAOs as a minimum, possibly mayors and Deputy Ministers) are brought together to discuss two or three key issues (a background paper should be issued one month before the meetings). This meeting should be about identifying challenges and working together toward pragmatic, effective solutions, not about fund raising, membership drives, or political advocacy. The approach may be somewhat like local Green Plans or Local Agenda 21s, common in the 1980s and 90s, but they should be customized by the community and should be annually updated. A minimum rolling, 10-year planning horizon, should be provided. These five (minimum) meetings and summary reports may well be the most important catalyst to focus common efforts toward key sustainability initiatives. They would also serve as an excellent early warning to key resilience challenges. The reports should be supportive to communities and agencies (not an audit) and should be an excellent first-stop for residents concerned with the transition to sustainability.

Cities and Sustainability—Statistics Canada

Statistics Canada should regularly provide consolidated, city-based data and establish a permanent webpage for (at least) each of the five major urban regions of Canada. The governments, businesses, institutions, and people of Canada should see this information as the most up-to-date, representative and accurate data on Canadian cities. As a minimum, the data should include population (with demographic and residency breakdown), GDP, all GHG emissions (Scopes 1, 2, and 3), energy and materials use, waste generation, health indicators, education levels and performance, and best approximations of populations, demographics, and residency status.

Evidence, science-based policies should be predicated on this data. People are entitled to their own opinions; however, the facts that drive civilization's narratives need to be common and readily available. Statistics Canada must remain one of the most respected sources of this information. The area for which the data pertains to, should however, as much as possible, coincide with the drivers of Canada's economy. The borders of these urban areas (larger agglomerations) will be consistent with the borders of local governments (a local government—municipality—is either in the area or not); however the overall area, as an urban system, will be defined by mobility, employment, and urban systems.[19] The borders of the urban areas, defined by mobility and work patterns, are like Conservation Authorities, where the area is defined to be consistent with watersheds.

---

[19] Two of the five urban areas may have contested borders. Toronto Region is proposed to be consistent with the Golden Horseshoe (including the GTHA and the outer ring with Kitchener-Waterloo, Peterborough, and Barrie). The Calgary-Edmonton area may be seen to be two unique urban areas separated by 250 km of mostly rural space. However, for both larger agglomerations, there is a benefit from a sustainability perspective, for the communities, and the rest of Canada, to optimize across the broader community and take a systems approach to the larger agglomeration.

7.14 Recommendations

International Mining and Materials Association

Like the International Energy Agency, the Toronto Board of Trade, and others, should catalyze the establishment of an International Mining and Materials Association (IMMA) in Toronto. IMMA should be tasked with assessing projected material demands and research on ways to reduce overall demand. Most mining and energy associations have dual mandates, production and supply, and often a distant-second objective, conservation. The IMMA should research ways in which the material needs of society, e.g., cities, can be reduced. For example, shifting to a conserver society and embracing greater circularity. Individual companies, or trade associations, will not promote new technologies like decentralized digital ledgers and electronic gold reserves (leave it in the ground) [53]. IMMA could emerge as the global authority on which potential source of materials has lower overall negative impacts on sustainability. The IMMA could help compare 'ethical oil' to low-carbon oil, or 'blood diamonds' versus 'conflict-free'. Canadian cities should support IMMA and encourage international peer communities to do the same.

Using the Power of Post-secondary Education

Canada has the OECD's highest share of 25–64-year-old population with post-secondary education (more than 55%). There are about 2.2 million students enrolled in Canada's post-secondary institutions, as well as more than 48,000 full-time teaching staff at Canadian universities [57], and an additional 72,000 academic staff, across 125 colleges and universities. In Canada's bigger cities, post-secondary students are the largest cohort of commuters, e.g., 600,000 per day in the Greater Toronto Area [58]. In at least Canada's five largest urban areas, joint academic degrees (common courses and programs) should be offered, starting with international students.

As Canada continues to encourage internationally trained doctors, nurses, and others, to immigrate to Canada, shortages of these key skills are exacerbated in other countries. The federal and respective provincial governments should work together to provide two academic spaces in similar programs for international students (either in Canada or a partnered low- and middle-income country university) for each graduate professional that immigrates to Canada.

Narratives, Not just Numbers

Canada's cities, especially the five larger metros, will determine much of Canada's future prosperity and influence the rest of the worlds. The people of these five (mostly) urban areas—the businesses, the employees, the residents, visitors, and children—need to know how the cities (metro areas) are performing, or at least have the information available to make a credible assessment. Statistics Canada can provide annual metrics for these communities (metro areas as CMAs); however, 'narratives, not just numbers' [59] are needed to inform on progress, provide encouragement, and not be about punishment. Statistics Canada can readily publish a suite of indicators for Canada's larger metro areas that mimic the sustainable development goals (SDGs) and capture the community's overall progress toward sustainability.

Data Management

The concept of subsidiarity is important in local government. This is especially the case in data security and management. The data management sector has undergone massive development over the last 10–15 years. Most of this was driven by the private sector, e.g., Google, Facebook, and the telecoms such as Bell, Rogers, and Telus.

Municipalities typically own the streets in cities; they oversee traffic management, snow removal, and waste management. They should also be a key part of data collection and security. Sidewalk Labs in Toronto (see Box 3.1) highlighted the need for trust, innovation, and professionalism in urban data systems, and what might happen in its absence. This will increase as the sustainability transition encourages mobility as a service and dynamic road pricing. Each Canadian residence should be equipped with a municipally supported tablet that provides energy and material flows information, emergency messaging, and community information.

An International Focus

There is no country of comparable size more dependent on immigration than Canada. Canada's strength is diversity, and cities are where this diversity is most felt. A steady supply of quality immigrants, and the continued ability to assimilate these newcomers, is not assured. Steps should be taken to maintain the country's international attractiveness and ability to welcome migrants. Some of these steps could include working to lower the costs of remittances (see Sect. 4.2), supporting the education of populations in low-income countries (see above), and providing hybrid and flexible visas for visitors to work and live in Canada. An additional way Canada could help convey its desire to be a good international citizen is to pursue with greater vigor the stated goal of meeting the UN's target of 0.7% of GDP to official development assistance (ODA, the current level is about 0.38% [60]). There is considerable discussion and international pressure for Canada to meet its NATO target of 2% of GDP spending on military spending. The targets, arguably two sides to the same coin, should be pursued with equal vigor (and innovation on methods to meet the targets).

7.15 Council's Climate Resolution

A professional engineer tasked with advising municipal council today on how to respond to Canada's climate changes expected over the next 50-to-100 years might provide a report along the following lines:

To: Municipal Council, Our Town, Canada.

From: Don Ho, P. Eng. City Engineer, Our Town, Canada.

WHEREAS combustion of fossil fuels such as oil and gas releases large quantities of CO_2, these and other greenhouse gas (GHG) emissions (about 57.1 billion tons in 2023) are leading to a rapidly changing climate (one of the fastest rates of change ever experienced in geological history, akin to a meteorite impacting the earth);

7.15 Council's Climate Resolution

AND WHEREAS the atmospheric concentration of CO_2 (and other greenhouse gases such as methane) was stable at 280 ppm since the Holocene age and our 'Goldilocks-like' climate began some 11,500 years ago, but concentrations rose with the Industrial Revolution starting around 1760 and passed the 'safe' 350 ppm level in 1989 and today is around 424 ppm and increasing about 2.4 ppm every year (the rate is still increasing—more than half the total CO_2 ever emitted through human activities occurred in the last 30 years). CO_2 emissions are expected to pass the 450 ppm level in 2040 at which point global temperature increase is on track to surpass 2.0 °C triggering climate 'tipping points' this century (such as loss of permafrost, Greenland ice sheet collapse, ocean circulation changes, boreal forest shifts);

AND WHEREAS Canadians, per person, are among the world's highest GHG emitters (and only partly because the country is big and cold), and the world is engaged in challenging negotiations to reduce GHG emissions and provide compensation through a Loss and Damage Fund (animosity and blame within and between countries is expected to increase);

AND WHEREAS Our Town has a 'net-zero' carbon by 2050 target, along with many other Canadian communities, and as legislated by the Government of Canada, however we are not on track to reach this target (no one is in Canada);

AND WHEREAS the GHG emissions that are under Our Town's direct control are less than 3% of the community's total emissions[20]; however community emissions, that are under everyone's influence, should account for 'Scope 3' emissions associated with consumption that occurs anywhere (e.g., air travel, cruises, and imported food that add at least an additional 35%). Each Our Town resident contributes over 20 tons CO_2 per year, not including their share of Canadian forest fires;

AND WHEREAS more than 2.5 billion people live in areas expected to be outside of a safe climate niche by 2060, and today more than 750 million people do not have access to electricity (more than 1 billion people living in low-income communities today are the most threatened by climate change, but have contributed virtually nothing to the problem);

AND WHEREAS most municipal infrastructure is designed with a factor of safety applied across a reasonably well-known climate horizon, such as rates of precipitation, temperature, and wind speeds, and that this stable planning horizon can no longer be relied upon beyond, say 2040;

AND WHEREAS Our Town, and related provincial and federal services, has an infrastructure backlog in excess of $3 billion (more than 30% of current value; about 25% of our infrastructure is operating beyond its projected lifespan);

AND WHEREAS the Government of Canada removed the consumer carbon tax on April 1, 2025;

---

[20] Canadian 'territorial' emissions are about 30% oil and gas sector, 22% transportation, 13% buildings, 11% industry, 10% agriculture, and 8% electricity (about 18 tons per person; not counting forest fires which were as much as an additional 18.8 tons per person in 2023).

AND WHEREAS climate change is only one of several interrelated planetary system threats, others include biodiversity loss, land-use changes; we require a systems or sustainability approach with dynamic solutions;

AND WHEREAS Our town, like other Canadian communities, is vulnerable to supply disruptions, such as medical supplies, food, and energy, and the electricity grid is in particular need of strengthening;

AND WHEREAS we are dedicated to reconciling with the First Nation communities that we share the land and increasingly the culture with, we recognize that work is still needed and maybe this task of saving mother earth may bring us closer together;

AND WHEREAS the following is broader than a standard municipal engineering recommendation, the integrated nature of the challenge, and the 'all hands-on deck' requirement for effective solutions, necessitates an approach beyond single professional disciplines and community mandates;

THEREFORE, BE IT RESOLVED that Council adopt the following ten-point sustainability plan.

1. Inform the community that we are in the danger zone. The climate is already changing, rates of global degradation are increasing, and massive change is underway but there is much we can do as a community to prepare, and to reduce the scale of the problem. At least once-per-year, residents should receive information outlining how their homes and our community can be made more resilient.
2. Recognize the built-in inertia (businesses striving for status quo, human reluctance to change, regional and international inequities) and be kind with each other in the challenges associated with the changes coming, while also being resolute that we are all in this together, so everyone needs to be part of the transition efforts.
3. Canada, because of its temperate climate and substantial water resources, including Our Town, will likely face significantly higher pressures to receive immigrants and international migrants. Housing flexibility, greater transience of citizens, and the need to build more durable partnerships with communities outside Canada should be anticipated.
4. The risk is high that related to this massive global change, there will be military hostilities in the next 50 years, including Canada. Our Town will of course be called on to support this war(s); however as a community we should assert our beliefs that a peaceful, supportive transition is possible, and preferable. We need to also show this within our own community.
5. Our Town's 'old world' infrastructure is rapidly moving outside state of good repair, with serious maintenance backlogs and rapidly declining resiliency. This is common with other, non-municipally managed, but equally critical infrastructure, such as electricity generation and distribution. A five-year rolling infrastructure management plan should be prepared for Council, with publicly discussed recommendations, links to other levels of government, and updated annually.

7.15 Council's Climate Resolution

6. Our Town municipality, businesses and residents, should anticipate the crossing of local climate and ecosystem tipping points. Lakes may become anoxic, greater fire intensity and frequency, flooding, and severe weather will be more common.
7. Our Town, being contiguous to That Town and Their Town, should forward this resolution to them and ask how we may establish a more durable sustainability partnership (e.g., common metrics, shared disaster response planning, coordinated economic development offices, mobility apps, and data management systems).
8. Integrated mobility (transit, ride sharing, bicycles, walking, and low-carbon vehicles) is one of the most important initiatives needed to reduce GHG emissions, enhance resilience, and increase economic development (while enhancing well-being). Our Town can work with neighboring communities, municipal associations, and provincial and federal governments to develop and implement and 'mobility as a service' programs along with supporting infrastructure.
9. Data collection, security, and communications systems are increasingly critical to the safe and sustainable operation of Our Town. Our Town needs to partner with residents, businesses and visitors, on data platforms, and the active monitoring and communication of sustainability metrics. A data management plan will be prepared and presented to Council for consideration. The principle of subsidiarity suggests that this plan be developed and implemented by local communities and scaled upward (across Canada and internationally).
10. Recognizing that we need to develop a more sacred relationship with the land as a key aspect in the transition to sustainability, we should seek a more durable partnership with our neighboring First Nations communities and residents of Our Town.

Possible Question and Answer Session Following the Report.

Q. What about geoengineering?

A. Geoengineering, the intentional effort to limit solar radiation from reaching earth, will probably be pursued within the next 50 years. Things are that bad. However, we will never be able to fine-tune impacts, and we do not yet know how effective interventions may be. There will be winners and losers. Whole countries may feel disenfranchised, and when we cannot come together as countries to establish preventative (and much cheaper) approaches to climate change, most people are very doubtful that we will be able to come together to manage a much more complex activity like geoengineering. Also, most engineers cringe that the term 'engineering' is applied to this approach as it is almost impossible to apply a factor of safety against the intervention.

Q. Why did your engineering predecessors not warn Council more loudly about climate change before?

A. Civil engineering curriculum has included climate change considerations for more than 40 years, and the Government of Canada, and other UN members are on record calling for substantive reductions to GHG emissions in 1987 when Our Common Future (the Brundtland Report) was released. Talking about planetary systems at a municipal council meeting is uncommon. Arguably our provincial and federal governments should be leading this discussion and implementing meaningful programs. However, the scale of the challenge and the changes required are so large, that some procrastination is inevitable, and partisan disagreements on possible approaches likely. Industry, especially the fossil fuel industry, is also still actively promoting a 'go slow approach' to safeguard profits as long as possible. Alternatives are not easily available. Canada is largely at the 'who will pay, and how much' stage. Local governments in Canada tend to be more pragmatic (and perhaps less partisan), and even though their mandated political remit is relatively small, their management of infrastructure is significant (municipalities in Canada own and manage about half the country's infrastructure). Municipalities in Canada will be tasked largely with safeguarding our infrastructure through the sustainability transition—that is why we are coming forward today with this resolution.

Q. Is not a tax on carbon a tax on everything? People are already facing affordability challenges, what you propose seems to imply future costs, much higher costs.

A. True CO_2 emissions are a significant by-product of our lifestyles, and in many areas easily available alternatives for businesses and residents do not yet exist. However, economists and policy experts tend to value pricing signals for their ability to nudge behavior. Canada will also be subject to border adjustment tariffs as other countries reduce their emissions faster than us. More than a quarter of the world's carbon emissions are already subject to a price on carbon. Canada is in a precarious position in that our emissions, per person, are among the highest in the world, and even though climate impacts will be severe in Canada, especially in the Arctic, our northern latitude is an enormous geographical advantage. We should expect global resentment toward us, as a country to rise. We are also starting to see the price of inaction on climate change, and from the perspective of Our Town, the costs to safeguard our infrastructure are higher than what the cost of prevention might have been.

Q. What should we tell our constituents, especially younger ones?

A. Things are changing, and the pace of change is likely to increase over the lifetime of anyone alive today. Climate change is just one symptom of our economic and social system being out of balance. The human species will get through this. Within 50 years our global society can transition to energy and materials levels that are sustainable. Getting there will not be easy, but it is much better than the alternatives. Also, the planet is very resilient and there is still much to safeguard. We can be optimistic.

Q. If we want a second opinion, who should we invite to Council?

A. Possibilities include Katharine Hayhoe, a Canadian climate scientist living in Texas, Christiana Figueres, who headed the negotiations for the Paris Agreement, and Corinne Le Quéré, an outstanding Canadian-British climate scientist at University of East Anglia.

Helpful books include Hannah Ritchie *Not the End of the World: How we can be the first generation to build a sustainable planet;* Gwynne Dyer *Intervention Earth: Life-Saving Ideas from the World's Climate Engineers;* Michael Mann *Our Fragile Moment: How Lessons from Earth's Past Can Help Us Survive the Climate Crisis*; Bill Gates *How to Avoid a Climate Disaster*; Minouche Shafik *What we owe each other: A new social contract for a better society*; Deb Chachra *How Infrastructure Works: Inside the Systems That Shape Our World*; and Parag Khanna *Move: How Mass Migration Will Reshape the World – and What It Means for You.*

References

1. Freud S (1930) Civilization and its discontents
2. Un-Habitat (2023) Urban migration
3. Bettencourt LM, Lobo J, Helbing D, Kühnert C, West GB (2007) Growth, innovation, scaling, and the pace of life in cities. Proc Natl Acad Sci 104(17):7301–7306
4. Dyer G (2024) Intervention earth: life-saving ideas from the world's climate engineers. Random House Canada
5. Waring TM, Wood ZT (2021) Long-term gene–culture coevolution and the human evolutionary transition. Proc R Soc B 288:20210538. https://doi.org/10.1098/rspb.2021.0538
6. Cline EH (2024) After 1177 BC: the survival of civilizations. Princeton University Press
7. International Energy Agency (2021) Global international fuel economy initiative
8. Chester M, Allenby B (2024) Infrastructure first principles for the Anthropocene. Environmental research: infrastructure and sustainability
9. Schodde R (2010) The key drivers behind resource growth: an analysis of the copper industry over the last 100 years. MinEx Consulting
10. Chachra D (2023) How Infrastructure works: inside the systems that shape our world. Riverhead
11. Bristow D, Kennedy CA (2013) Urban metabolism and the energy stored in cities: implications for resilience. J Ind Ecol 17(5):656–667
12. Hoornweg et al (2025) A day in the life: energy and material flows in Montreal, Toronto and Vancouver (In preparation)
13. https://www.reuters.com/business/aerospace-defense/air-travel-hit-by-global-cyber-outage-2024-07-19/
14. https://www.cbc.ca/news/politics/rogers-outage-human-error-system-deficiencies-1.7255641
15. Government of Canada (Accessed 11–3–2024) https://www.getprepared.gc.ca/cnt/rsrcs/pblctns/yprprdnssgd/index-en.aspx
16. Clark J, Straus SE, Houston A, Abbasi K (2023) The world expected more of Canada. bmj, 382
17. https://ncc-ccn.gc.ca/events/weekend-bikedays (Accessed 11–3–2024)
18. https://web.archive.org/web/20090902004409/http://www.downtownwinnipegbiz.com/home/events/ciclovia/ (Accessed 11–3–2024)
19. City of Toronto (2024) Toronto BIA office
20. Charenko M (2015) A historical assessment of the world's first business improvement area (BIA) the case of Toronto's Bloor west village. Canadian J Urban Res 24(2):1–19

21. Stewardship Ontario (2013). The story of Ontario's Blue Box.
22. UNFCCC (2023) UN Climate Change Conference – United Arab Emirates
23. Silver D, Taylor Z, Calderón-Figueroa F (2020) Populism in the city: the case of ford nation. Int J Pol Cul Soc 33:1–21
24. Kiss SJ, Perrella AM, Spicer Z (2020) Right-wing populism in a metropolis: personal financial stress, conservative attitudes, and Rob Ford's Toronto. J Urban Aff 42(7):1028–1046
25. Applebaum A (2024) Autocracy, Inc: the dictators who want to run the World. Random House
26. Hemingway E (1926) The sun also rises. Modern Library, New York
27. Lopez BH (1990) The rediscovery of North America. First appeared in Orion magazine 1990
28. Merriam Webster Dictionary (Accessed 12–5–2024)
29. Hemingway E (1932) Death in the afternoon. New York, London, C. Scribner's Sons
30. Lopez BH (1990) The rediscovery of North America. Penguin Random House
31. Leopold A (1949) A sand county almanac. Oxford University Press, New York
32. HFS Books, The Rediscovery of North America, by Barry Lopez. University Press of Kentucky
33. Kuwae M, Yokoyama Y, Tims S, Froehlich M, Fifield LK, Aze T, Tsugeki N, Doi H, Saito Y (2024) Toward defining the Anthropocene onset using a rapid increase in anthropogenic fingerprints in global geological archives. Proc Nat Acad Sci 121(41):e2313098121
34. Meadows DH, Meadows DL, Randers J, Behrens WW (1972) The limits to growth-club of Rome
35. Supran G, Rahmstorf S, Oreskes N (2023) Assessing ExxonMobil's global warming projections. Science, 379(6628)
36. Steffen W, Richardson K, Rockström J, Cornell SE, Fetzer I, Bennett EM, Biggs R, Carpenter SR, De Vries W, De Wit CA, Folke C Sörlin S (2015) Planetary boundaries: Guiding human development on a changing planet. Science, 347(6223):1259855
37. Investors for Paris (2024) Playing with fire: Canadian Insurers and Fossil Fuels
38. Ritchie H (2024) Not the end of the world: how we can be the first generation to build a sustainable planet. Random House
39. Greenspoon L, Krieger E, Sender R, Rosenberg Y, Bar-On YM, Moran U, Milo R (2023) The global biomass of wild mammals. Proc Natl Acad Sci 120(10):e2204892120
40. Ledger SE, Loh J, Almond R, Böhm M, Clements CF, Currie J, Deinet S, Galewski T, Grooten M, Jenkins M, Marconi V (2023) Past, present, and future of the Living Planet Index. npj Biodiversity, 2(1):12
41. https://www.cbc.ca/news/business/kleenex-canada-1.6947410 (9–9–2024)
42. CBC News (2024) Canada's rural communities will continue long decline unless something's done, says researcher. https://www.cbc.ca/news/canada/london/immigration-rural-canada-1.7106640#:~:text=Since%202016%2C%20rural%20Ontario%20has,per%20cent%20increase%20in%20seniors.
43. Iveson D, Eidelman G Toward the metropolitan mindset: a playbook for stronger cities in Canada.
44. IEA 2023 and Wood Mackenzie 2024
45. Rushdie S (2021) Languages of truth: essays 2003–2020. Random House
46. Economist Intelligence Unit (2024) Global economic outlook
47. Boeree L (2023) https://forum.effectivealtruism.org/posts/EPB3kSwEAx6HYJNaA/ted-talk-on-moloch-and-ai (2024–09–14)
48. Brown L (1978) The twenty ninth day. Norton and Company
49. Government of Canada (2023) Notice—supplementary Information for the 2024–2026 Immigration Levels Plan
50. McAuliffe M, LA, Oucho (eds) (2024) World migration report 2024. International Organization for Migration (IOM), Geneva
51. Institute for Economics and Peace (2020) Ecological Threat Register
52. Luft J, Ingham H (1955) The Johari window, a graphic model of interpersonal awareness. In: Proceedings of the western training laboratory in group development
53. Pollon C (2024) Pitfall: the race to mine the World's most vulnerable places. Greystone Books
54. Johnston D (2018) Trust: twenty ways to build a better country. Signal

References

55. Lewis, M (2025) Who is government?: the untold story of public service
56. Marceau R, Bowman CW (eds) (2012) Canadian academy of engineering, vols I&II . Winning as a Sustainable Energy Superpower, Canada
57. Statistics Canada (2023) Released 2023–11–01
58. StudentMoveTO (2022)
59. Shore C, Wright S (2024) Audit culture: how indicators and rankings are reshaping the world. Pluto Press.
60. OECD (2024) Development co-operative profiles

Open Access This chapter is licensed under the terms of the Creative Commons Attribution-NonCommercial-NoDerivatives 4.0 International License (http://creativecommons.org/licenses/by-nc-nd/4.0/), which permits any noncommercial use, sharing, distribution and reproduction in any medium or format, as long as you give appropriate credit to the original author(s) and the source, provide a link to the Creative Commons license and indicate if you modified the licensed material. You do not have permission under this license to share adapted material derived from this chapter or parts of it.

The images or other third party material in this chapter are included in the chapter's Creative Commons license, unless indicated otherwise in a credit line to the material. If material is not included in the chapter's Creative Commons license and your intended use is not permitted by statutory regulation or exceeds the permitted use, you will need to obtain permission directly from the copyright holder.

Chapter 8
Thoughts of Others

Cities bring people together, in-person and virtually. In preparing this book, interviews were held with about 30 people active in the urban sector, half in Canada, and half international.[1] A key takeaway in talking to these individuals, and others, is how open and engaged people are on the issue of cities, and how hopeful and keen they are to help. There is enormous latent support for the transition to sustainability.

City practitioners and researchers are both terrified and hopeful. The trends are alarming and impacts on cities are already manifest from a rapidly changing climate, growing inequity, angst among voters and residents, and massive technological

---

[1] The following people were interviewed in preparing this book: Judy Baker, Alan Broadbent, Jeb Brugmann, Harriet Bulkeley, Donna Chiarelli, Greg Clark, Billy Cobbett, Bharat Dahiya, Andy Deacon, Sybil Derrible, Ursula Eicker, Jane Engel, Warren Evans, Jon Fink, Mila Freire, Giles Gherson, Anne Golden, Jacquie Hoornweg, Vijay Jagannathan, Abhas Jha, Chris Kennedy, Dave Layzell, Jennifer Lenhart, Erin Mahoney, Patricia McCarney, Elizabeth McIsaac, Lisa Prime, Mary Rowe, Richard Stren, Jag Sharma, Enid Slack, Marilyn Spink, Raf Tuts.

shifts. Those who monitor the trends are aware how much of the impact originates from urban lifestyles, and how cities will need to significantly increase resilience in adapting to the coming climate and social changes.

Civil engineering; energy and material flows; public health; transportation; infrastructure development; finance; economics; political science; and other social sciences—the disciplines of cities are many. So too the numbers of people working in cities either as specialists or general managers. And everyone working in the field of sustainability is acutely aware of how cities and urban systems underpin efforts toward sustainability.

Cities are bracing as geopolitics and the rise of autocracy and nationalism seem to follow in pathetic fallacy with growing global climate shifts. Cities being immobile and supported by myriad supply chains, as well as undergoing massive growth in population, and soon for some, marked declines in population, are particularly vulnerable to the rising turbulence. However as outlined in Chap. 7, cities will endure. They are also humanity's greatest hope.

In interviewing people for this book, a few common themes emerged and a handful of intriguing inputs.

One of the best ways to learn about Canadian cities is talking to those who observe them from outside the country. A perceived strength of Canadian cities is undoubtedly their ability to accommodate new residents. Diversity and tolerance are seen as powerful strengths, perhaps the superpower of Canadian cities.

Urban practitioners, closer to the 'coal face' of day-to-day urban service delivery, are however generally less optimistic about Canadian cities. Perhaps this reflects their ability to discern trends earlier than most. Congestion, urban finance (considering the overall economy), declining service levels, insufficient and aged infrastructure, fracturing governance, and declining civility were identified as threats.

Like the economics discipline, there are macrourbanists and microurbanists, each focusing on different attributes of the city. Macrourbanists are particularly wary of the increasing speed of climate change (and other earth system changes), the likely human migration that will follow, and how cities will together, or not, work to address the changes, and geopolitical ramifications.

Few believe the energy transition will meet the Paris Agreement aspirational limit of 1.5 °C, or even 2.0 °C, global temperature increase. Most urban practitioners are bracing for global temperature increases above 2.5 °C, although they typically believe modest impacts will be experienced over the next 20 years. Local and global ramifications of these temperature increases are not yet well understood.

Cultural diversity, relative levels and service standards of urban infrastructure, and attractiveness to potential immigrants (although declining) were raised as Canadian urban strengths. Key threats included housing affordability, sustainable finance and governance, increasing populism, and challenges to effectively communicate with people of the community. Cybersecurity and data management (with analytics) and improved transportation were raised as priorities.

Macrourbanists are focused on perceived and real declines in civility, the need for local resilience and system hardening, and a much greater emphasis on equity and inclusiveness. Enhanced appreciation of cities as the commons and how to accommodate future migrants was raised by several Canadian researchers. One of the most expressed hopes by those interviewed living outside Canada was the need for Canadian cities to continue being 'among the most livable cities while accommodating new migrants.' A common concern raised by people residing in Canada is the limits of local government's ability to govern and finance within the current framework of local, provincial, and the federal governments.

Several commentators expressed deep challenges and pessimism, especially in national governments and the apparent (hopefully temporary) erosion of civility and an appreciation of the common good. The need for cities to reinvigorate engagement and collaboration was raised by many.

Urban governance and management were thought to "not be a competition", but "they are competitive". How collaboration might emerge and how a more durable partnership between local government and the community might evolve was raised. A more collaborative government is needed that can tolerate mistakes and include a longer view in its deliberations.

A rough categorization of comments into challenges (and pessimism) and opportunities is as follows.

8.1 Challenges

Most people interviewed shared a sense of urgency and a strengthening need to improve civility and prepare for large-scale changes in climate and social norms. The growing impact of urban heat islands, along with more climate-driven disasters (within Canada and globally) was raised. There was recognition of the ongoing urban–rural conflict, as well as the growing divergence between the priorities of men and women.

Cars and cities do not mix well—cities that break free from reliance on automobiles provide markedly better livability (this is not yet apparent in Canadian cities).

The urgent need for a systems level analysis on urban services was raised as well as how technological advances are important. Behavioral shifts however trump technology. One interviewee believed that collaboration needs to be incentivized, e.g., more application of ecosystem services, life cycle costing. Conservation Authorities that service regional ecosystems should be re-invigorated. Similarly, regional development agencies need to be encouraged (within existing governance structures).

Several people opined that local governments need to be empowered as the key level of government to own data and its flow. Digital twins and remote sensing (and AI) need a local government perspective and management capacity.

Transportation, mobility of people, supply of energy and materials (and waste removal) remain a critical urban service for all governments. Montreal's progress was restrained by political interference, e.g., sovereigntist aspirations. This appears to be continuing.

Canada was recognized for good governance but not good local government.

Many developing countries increased incomes through manufacturing, e.g., China. This opportunity is less available for today's low-income countries. Exporting service jobs is likely, e.g., nursing. This will impact Canada. Global supply chains are also threatened, and re-shoring has significant limitations.

The 2.5 billion new urban residents expected by 2050 were highlighted and how they are expected to place particular strain on African and South Asian cities (both regions especially vulnerable to climate impacts).

Canadian cities and Canada are seen to be 'losing our shine'. There is concern about a rising informal economy in cities. Local governments especially need to develop policies to share intellectual property (and data systems). Urban sprawl was facilitated in Canada because of cheap energy (gasoline, electricity, fossil gas). As vehicle and housing size (area per resident) grow, the net-zero energy transition needs to be accompanied with an overall reduction in material use as well.

Globally there is a noticeable 'fragmentation of voice'. A new city voice is emerging (an orchestra with no conductor).

The degree to which it is increasingly easier to break than make things was raised. A key challenge identified was the limits of emergency management in preparing for multiple and cascading events.

Limited self-governance and financing powers of municipalities in Canada were raised. So too, poor integration of educated people and professionals into Canadian cities. Weak urban–rural linkages were also highlighted. This was also linked to continued, perhaps growing, frictions between provinces and the federal government, and stifling of local government innovation and financial growth.

Housing availability and affordability, and a few innovations on new living arrangements were discussed. So too continued local government fracturing (the parts not reflecting on the sum of the whole).

The need to convene stakeholders to address key priorities was identified (e.g., annual meetings with no more than 2–3 priority areas, focus on specific large cities, inclusion of all government levels and representative businesses).

The significant demands for Canada to spend to NATO's two percent of GDP target, yet almost no mention of meeting Canada's ODA target (0.7%) and climate finance commitments under UNFCCC[2] was observed.

---

[2] Canada is one of the 24 countries that committed in 2009 to contribute at least $100 billion per year to low-income countries to help them transition to low-carbon economies and adapt to a changing climate. At COP 29 (2024), this was raised to $300 billion, but still considered woefully inadequate by low-income countries (who demanded $1.2 trillion). The prospect of the USA withdrawing from the Paris Agreement would further impact these negotiations.

8.2 Opportunities

Immigration and Canada's openness were raised by many as an opportunity. In Canada, immigration is what you do, not who you are. Hybrid, or 'golden' visas may emerge as a key enabling feature of global cities, especially those that may serve as 'hub cities', e.g., Singapore and Dubai (Toronto?). Vancouver could emerge as a gateway to Asia (like Miami to Latin America).

Climate change could drive professionalism and collaboration. Collaboration is increasingly more important than competition and may also be driven by climate change. The good fortune of Canada's geography was identified by several interviewees.

Canada's ability to emerge as an 'energy superpower' was raised and may be facilitated by pursuing an approach where new (low carbon and clean) electricity and hydrogen generation are shared between domestic use and export. This was seen to be linked to how cities concentrate people as well as minerals (suggesting the need for a complementary approach to urban provisioning and new mining for critical minerals).

The rising role and of secondary cities was identified, as was the potentially helpful voice of mayors, e.g., Paris, Barcelona, and Montreal. The importance of institutions is growing—this is a relative strength of Canada and Canadian cities, e.g., ICLEI, C40, UN-Habitat.

A team Canada approach is needed to address Canada's loss of mining stature and reindustrialization that supports just-in-time manufacturing (and robust supply chains).

Growing inequality needs to be staunched; the benefits of immigration need to be communicated by additional community spokespeople and agencies. A shift in the nature of property ownership and the rights of nature is underway. So too, a shift away from the sole power of the nation-state. This will be accompanied with bioregions, urban regions (megaregions), and a shift to indigenous governance practices.

Northern Canada was identified as a key emerging global region.

A shift to pricing vehicle-kilometers-traveled that has the potential to reduce congestion and improve local government financing was suggested.

A seven-generation perspective supports greater consideration of the earth systems acting on cities, e.g., volcanology and seismology in the Cascadia Corridor, long-term ecosystem health of the Great Lakes. The latent culture of some communities was raised, e.g., San Francisco and then Portland banning facial recognition.

Significant capital accumulation is anchored in Canadian cities. This may be useful when linking climate challenges to finance and capital appreciation (wealth generation).

Open Access This chapter is licensed under the terms of the Creative Commons Attribution-NonCommercial-NoDerivatives 4.0 International License (http://creativecommons.org/licenses/by-nc-nd/4.0/), which permits any noncommercial use, sharing, distribution and reproduction in any medium or format, as long as you give appropriate credit to the original author(s) and the source, provide a link to the Creative Commons license and indicate if you modified the licensed material. You do not have permission under this license to share adapted material derived from this chapter or parts of it.

The images or other third party material in this chapter are included in the chapter's Creative Commons license, unless indicated otherwise in a credit line to the material. If material is not included in the chapter's Creative Commons license and your intended use is not permitted by statutory regulation or exceeds the permitted use, you will need to obtain permission directly from the copyright holder.

Chapter 9
Conclusion, A Call to Action

Chances are, today's unborn child, if given the chance, would choose Canada as one of the countries to be born in. Canada is one of the world's most inclusive and arguably, hopeful societies. This optimism, however, is waning. For example, Canada fell another three places in the 2025 World Happiness Report (18th place in 2025, 15th in 2024, and 5th in 2015) [1]. The significant variance in levels of happiness between Canada's young and old, remained among the largest in the world.

Canada won the geography lottery and is one of the best countries positioned to deal with climate change, in addition to being replete with freshwater, agricultural lands, and resources like critical minerals, potash, and all forms of energy. Located next to the world's largest economic and military power, Canada also benefited geopolitically for the last couple of centuries. This relative strength is now in question however with the US Administration's growing nationalism (and isolationism). Politics and the permeability of the Canada-US border may change; however geographic proximity will not. Communities along the Canada-US border will likely always be well connected.

Canada's geographic attraction was not always shared. Jacques Cartier in 1534, when first sighting the north shore of the Gulf of St. Lawrence quipped "The land God gave to Cain" [2]. However, for the last 100 years, Canada was reasonably well governed and largely grew comfortable with its place in the world. Canada's cities consistently place in the top tier for livability. People still want to live in Canada, especially in Canada's flourishing cities.

Canada's relatively bright future also comes with a dark aspect. Per person, Canadians inflict more damage to earth systems than any other country. Canada's cities are among the most energy- and materials-intensive ever, anywhere. One Canadian has the materials metabolism of more than 12 elephants. Canadians helped drive the *Great Acceleration*, at times even getting out and pushing. Canadians do not travel lightly on the planet.

Canada's impact on the planet is driven by two broad categories. Canada grew with a resource mindset. Beginning with the export of beaver pelts and timber, and abrading at anything—First Nations, environmental safeguards, human health, equity—that might limit exports and profits. Energy conservation, for example, was never taken seriously in Canada, as new resources could usually be developed. In most categories, Canadians use about twice as much energy and generate about twice as much waste as the average person in a high-income country.

Canada welcomes the resource industry with open arms. More than half the world's mining companies are listed on the TSX, and Canada remains the largest supplier of oil and gas to the USA. The resource industry has, however, left the country with significant unfunded liabilities. Starting with one of the world's first petroleum wells near Oil Springs, ON in 1858, there are now more than 300,000 abandoned petroleum wells in Canada (likely requiring over $350 billion to remediate). One of Canada's first mines started near Trois-Rivieres, QC in 1737. There are now more than 10,000 abandoned mining sites across the country (requiring more than $250 billion to remediate).

Canadians might be surprised hearing that despite its extensive mining industry, the country is a net-importer of minerals since 2006. The resource and energy demand of Canadian lifestyles are enormous. Canada's cities are some of the most automobile-dependent and least densely populated in the world (~2300 p/km^2 along with the USA and Australia, compared to ~ 6700 p/km^2 in Europe and Japan, and ~ 13,600 p/km^2 in most developing countries). Some large cities in Canada still do not have residential water meters. Canada's vehicle fleet is the largest (by weight and size) in the world. Canada is the only advanced nation that has seen greenhouse gas emissions rise since 1990.

Over the next 100 years, Canada, especially Canadian cities, will be 'ground zero' for the *Great Deceleration* of shifting resource and energy consumption (aka the 'great simplification' [3]). Canada can stay in the car, or again, get out and this time help pull.

The Great Deceleration will manifest in myriad ways. For example, Brexit, the Trump Administration's imposition of tariffs (with a national annual account balance, deficit, more than $1 trillion), China's rush to advanced manufacturing, western Canada's growing alienation, are symptoms of massive shifts in demographics and planetary systems. Cities drive these shifts and are typically most impacted by them.

Canada's cities can take pride in many firsts. Ottawa first closed streets to cars in 1970, encouraging bicycles on Sundays. Waste separation and blue box recycling were launched in Kitchener in 1983. The world's first business improvement area (BIA) started in Toronto's Bloor Street in 1970. These ideas were replicated around the world. For example, Bogota's renowned *cyclovia* began in 1974, and there are now more than 4500 BIAs around the world.

Canada also helped launch and strengthen many of the world's most important institutions. Canada is a full and active partner in the global family of nations. This experience is needed again, as the world requires strengthening of existing institutions (or sunsetting), and new methods of global cooperation.

One of the most striking aspects of Canada's last 100 years is the country at war. Canada's role in WWI and WWII were seminal to country's place in the world. Canada often may feel it is in the shadow of the USA, especially in recent years.

However, the similarities between people residing in southern British Columbia and Ontario with those of near-by Washington and Portland, or New York and Michigan states, are likely greater than between those of US residents living in the North and South. How the communities of Cascadia and the Great Lakes Region develop over the next 50 years will determine much of the future of both Canada and the USA.

Since 1920, 122 new countries emerged or re-emerged from colonial rule. The initiating charter of the UN was signed in 1945 by representatives of just 50 countries. Twenty-two countries participated in the 1912 Olympic Games, while 206 countries participated in recent Olympics. Canada and the USA are unique in that they saw the peaceful consolidation of territory over the last century (HI, AK, and NL). Arguably, a closer shared living arrangement between Canada and the USA, and Mexico, across North America, would benefit everyone.

Today Canadians do not feel as if they are living in the country that had the world's fastest-growing economy for the last 100 years (Canada's GDP grew more than the USA, or countries in Europe). This disillusion is most severe with the young. The 2024 World Happiness Report found Canada's old (60+) ranked their subjective well-being (SWB) 8th globally, one of the world's highest. However, Canada's young (< 30) ranked their SWB 58th globally: the largest difference among the 71 countries surveyed.

Canada's share of the global economy peaked in the 1970s and is on track to continue to decline this century (a function of relative decline in population and productivity). This contributes to the waning happiness of Canada's youth (along with degrading planetary systems). The relative shift in economic fortunes, along with population shifts, is also leading to growing regional discontent.

Canada's cities will need to mediate much of this discontentment and possibly temper provincial dictates. For example, Montreal, by far the largest (and growing) contributor to Quebec's economy, may have less desire for separatist aspirations and language requirements. So too Calgary and Edmonton may grow closer together, aspiring for more global representation and emphasis on technology, while diverging from the province of Alberta's greater emphasis on oil and gas development.

A country as geographically diverse, and with as many newcomers as Canada, will always have urban–rural, and regional tensions. Canada's cities (as CMAs) will however likely need to take on a more active role in addressing, and ameliorating, these tensions. Especially as future events are likely to demand much greater adaptive capacity in cities that are home to more than 80% of Canada's wealth generation and population.

Canada's cities—as a minimum, the largest agglomerations (Montreal, Ottawa-Gatineau, Toronto Region, Calgary-Edmonton, Vancouver), and ideally the next largest CMAs of Quebec City, Winnipeg, London, Halifax, Windsor, Victoria, and Saskatoon (with populations over 300,000), plus hopefully the remaining 20 CMAs—need to work together. This will not be easy as Canada's cities are among the world's most fractured and contested. Asking Calgary and Edmonton to work as one, along with Red Deer, is daunting. Or Toronto Region, for example, the natural urban system in need of service planning and optimization, is made up of

106 local (municipal) governments with 1200 municipal politicians, 34 transit agencies, 17 electricity distributors, 25 school boards, 8 health networks, 70 chambers of commerce, and 25 publicly funded colleges and universities with more than 40 campuses.

Canada's largest city, Toronto Region, makes up 70% of Ontario's economy, and almost a third of Canada's, and yet there is no politician or professional employee, at any level of government, who speaks for the whole community. The whole is largely at the mercy of the parts. This is a similar challenge with Canada's other large cities. Cities, the large messy complex systems that they are, need to operate with a metropolitan or regional mindset, and even more importantly, a sustainability mindset. Sustainable development in Canada, and the rest of the world, will grow from the ground up.

Chapter 6 presents the halftime report for seven generations of Canada's cities (1920–2120). Tables 6.1 and 6.3 highlight key trends in urban Canada, and possibly for the country, and rest of the world over the next several decades.

With climate change and the likelihood of 3 billion people being exposed to unprecedented, unbearable heat, migration, and sustainability will be inextricably linked. Geoengineering will be aggressively proposed. So too, far greater demands for climate reparations by the world's low-income countries.

The Johari window, with its known knowns, known unknowns, and unknown unknowns, applies dramatically to Canada's cities. The known challenges are understood within ranges. Estimating the magnitude and timing of the peak is the challenge. How Canada deals with an aging population, peak human population (likely 10.4 Bn around 2080) will drive peak energy demand and waste (and peak GHG emissions). The deceleration in population growth is already being felt in many countries. Canada's population, and the vitality of cities, is dependent on the continued ability to welcome immigrants.

Known unknowns for Canada's cities include Canada's continued willingness, and ability, to accommodate new immigrants, possibly with hybrid visas and residency arrangements. How geopolitical frictions manifest in Canada's cities, bringing about social unrest and disrupted supply chains, is not yet known, but is something expected. Shifts to technologies and social norms are also not yet known but anticipated. Some, like leaving gold and oil in the ground, are possible, but not yet likely. Canadian cities might also benefit as places of relative peace to attract global creatives and intellectuals as they leave their communities of greater conflict.

The collection and application of data, artificial intelligence, the space industry, autonomous vehicles, and new pharmaceuticals and gene sequencing are technologies that will force cities to adjudicate between rolling out the welcome mat or closing doors. And above all, communities need to engender trust, for trust is the glue of civilization. And trust grows most quickly in urban areas.

Chapter 7 makes several recommendations for Canada's cities to meet the triple planetary crises while building a nurturing and prosperous society at home and abroad. These include (i) new approaches to housing; (ii) appointment of Chief Resilience (or well-being, sustainability) Officers to at least Canada's five larger urban agglomerations (broader than individual municipalities); (iii) Statistics Canada to provide annual data for at least the 20 largest CMAs (and collectively for the five larger urban agglomerations); (iv) as GDP data is regularly provided by Statistics Canada for the five larger agglomerations, provide at least 1% of the HST directly to local municipalities (provided their annual GDP increase was larger, and GHG emissions declined more than the national average, provinces should also contribute); (v) establish an International Mining and Materials Association in Toronto; (vi) in at least the five larger agglomerations, the federal and provincial governments should coordinate joint academic degrees (common courses and programs), starting with international students (within five years, for every professionally trained immigrant entering Canada, e.g., doctors, nurses, engineers, Canada should train at least two international students with similar degree programs in Canada, or in their country (region) of origin); and, (vii) municipalities (subsidiarity to local governments) should be supported by a nationally appointed Privacy Commissioner and retained expertise in the Ministry of Science and Technology, to develop public-interfacing place-based data systems. The pace of change is increasing in most sectors. Governments, at all levels, utilities, and service agencies, should provide longer-term strategies (e.g., five-year plans), with annual updates. For example, GHG mitigation targets, or productivity increases, at a local level should have annual targets and progress reporting.

One 'small' but important effort that should be aggressively pursued is to bring down the cost to send remittances from Canada. The cost remains high (6.5% which is higher than Australia, France, Germany, Italy, Korea, Saudi Arabia, UK, or USA, and well-above the G20 target level of 5%, and more than twice the SDG target of 3% by 2030). Supporting immigrants to help their remaining families back home (through remittances, educational support, etc.) is good for Canada and good for the rest of the world.

Two narratives are key to the future of Canada's cities: integrating migration and sustainability, and balancing scarcity and sufficiency.

In the next few decades, Canada's communities will renegotiate the relationship between each other, regions, provinces, the federal government, and probably the international community. Financial architecture will be revised, probably with a larger role for local revenues and participation. This transition will involve First Nations, future generations, rich and poor. Good managers, good leaders, the entrepreneurial spirit, and new businesses, residents, and visitors—all are needed.

As complex emerging systems, cities bring people, energy and materials together, they generate new ideas and send out wealth and waste. Cities are where our better angles are given free reign, and where they meet the devils of the details. Multilateralism is messy, governance is muddled, and managing most cities is complicated. Yet, like how fractals replicate in scale from a single twig to a mighty forest, what happens in Canadian communities replicates outward. Cities are where our stories originate, and where those stories that capture dreams and ideals are applied.

In 2020, 366,942 babies were born in Canada [4], another 134,353,058 babies were born in the rest of the world [5]. Of these 767,000 children that are likely to eventually call Canada home, 111,215 [6] are expected to reach their 100th birthday [7]. In 1920, there were likely only a few, if any, centenarians in Canada. In 2020, there were 8546 Canadians aged 100 years and older. In 2120, this number is expected to increase at least 13-fold, providing a hopeful indication for the country's future in a turbulent world.

Building a hospitable Canada for children being born today, in a world that will be safe and flourishing in 2120, is a daunting task. Toronto was first settled by the Iroquois in the village of Teiaiagon near today's Humber River. The Constitution of the Iroquois Nation, believed to date back to 1142 when a total eclipse occurred in the region, recommends consideration for seven future generations (about 200 years) in decision-making [8].

The halftime report for seven generations starting 1920 would be mixed. Degradation of planetary systems is severe, and peaceful coexistence within Canada and around the world appears to be declining. Trends suggest growing conflict and crossing of several major tipping points in the next few decades. On the other hand, there is much cause for optimism. Global populations will peak and markedly decline within the next 100 years. The transition will be difficult, but overall degradation of planetary systems should decline with stabilizing populations.

Carbon-free energy widely exists and is now often cheaper than fossil fuels, and likely to be broadly available before 2080. Reconciliation with peoples and planetary systems is still possible and probable.

Canadians are still benefiting from the decisions of those who in 1920 built a prosperous country with vibrant cities and a desire to be a positive influence in the world. The next 100 years are expected to be more challenging to navigate. Canada and its cities are far from perfect, yet there is likely nowhere better, where people can come together to build a home for today's children and the generations that follow.

A Call to Action

Assessing Canadian cities, over a period of seven generations, benefits from a geological perspective that recognizes uncertainty and requires humility, especially for the second 100-year-half yet to come. Looking back, discontinuities, or hinge moments, in social stratigraphy are evident. The 1931 Statute of Westminster afforded Canada full political independence. In 1970, the October Crisis and martial law gripped Quebec, following in 1971 Toronto overtook Montreal as Canada's largest city. Canada's involvement in WWI and WWII and the Afghanistan conflict, and the country's role in developing and encouraging multilateralism, were formative. Illustrating a globally unique form of federalism, along with ten provinces, Nunavut emerged in 1999 as Canada's third territory, the first to be overseen by an Indigenous government. The resource industry was particularly powerful in shaping cities like Calgary, Edmonton, Saskatoon, Sudbury, Saint John, and service centers like Vancouver (mining and forestry), Toronto (mining and the TSX), and Montreal (Hydro Quebec). COVID-19 and its impact on Canada will have implications for decades.

In 1947, the partitioning of India drove the largest mass migration in human history. The newly created United Nations voted to partition Palestine, and the Cold War was dividing the world into opposing camps [9]. The reverberations of these fractures and tectonic shifts are still being felt today [10]. Canadians sense we are entering another hinge moment. The immobility of cities within fluid world events is having profound consequences on the country's economy, governance, well-being, and soon, a changing role for cities.

Looking forward, these things we know. President Trump and the US shift to isolationism will have lasting changes on Canadian cities. Climate change will intensify its impact on ecosystem services. Cities will need to dramatically increase resilience and adapt to growing numbers of climate migrants. Canada and its cities will also need to address global calls for reparations from countries with far fewer financial resources that contributed almost no greenhouse gas emissions (the world's largest market failure, ever). Global debt, for example Canada's federal deficit of more than $60 billion in 2024, and the US deficit of $1.8 trillion (6.4% of GDP), has enormous implications for cities. So too calls to increase military spending beyond 2% of GDP.

There is growing recognition of the polycrisis (aka metacrises) facing humanity. Interconnected, amplifying, systemic risks have emerged as intractable global challenges. Cities, as collective communities, or urban agglomerations, are sensing these changes most intensely as they outlive local and global businesses, and most likely their national governments. In today's connected and interdependent world, Calgary cannot succeed long term if Kinshasa and Jakarta, or Halifax, fail. The scale and pace, and the means of support, as well as the degree of isolation, will ebb and flow through social input; however, no country, or city, can wall itself from a world of change.

A defining characteristic of Canada and Canadian cities over the last 100 years is the degree of permeability across borders. The US-Canada border was one of the world's most permeable international borders. Before 9–11 (2001) Canadians and Americans could cross the border with nothing more than a driver's license. The border has thickened, especially recently with the second Trump Administration; however, with regions like Cascadia and the Great Lakes Region, a relatively permeable international border remains possible and in both countries interest. Canada's provincial borders, on the other hand, are among the world's most impermeable. Interprovincial trade barriers are significant, and resource development with provincial ownership suboptimizes Canada's economy (Sect. 7.7).

Canadians are ardent supporters of multilateralism but rarely view the merits of this cooperative approach to governance within the country's urban agglomerations. Canada's urban regions are among the world's most fractured and are hindered by the lack of wholistic, cooperative, action. The challenges of these urban areas are exacerbated by the complete provincial control over municipalities granted by the country's constitution. Multilateralism, over the last 100 years, was mostly between countries. Canada, in leading multilateralism over the next 100 years, will need to highlight how to cooperate across different levels of governments and institutions. This cooperation needs to be nurtured within urban agglomerations, e.g., across Toronto Region, as well as between cities, e.g., linking Montreal and Boston, as well as Montreal and Kinshasa.

Three other areas require Canadian city leadership: life expectancy and well-being, civility, and pragmatism.

Life Expectancy and Happiness

In 1800, no region in the world had a life expectancy higher than 40 years. In 2021, the global life expectancy was amazingly over 70 years (82 years in Canada). In 1900, the average life expectancy of a newborn was just 32 years. Globally, by 2021 this more than doubled to 71 years [11]. In 1921, infant mortality in Toronto was a staggering 136 deaths in the first year of life per 1000 births. Today the rate is down to about 4 deaths per 1000 births.

One of humanity's greatest accomplishments over the last 200 years is the increase in life expectancy. This remarkable achievement was brought about by advances in medicine (e.g., vaccines), public health (e.g., water and sanitation, reduced air pollution), and improved living standards.

Looking for a broader measure of human progress, Professor Helliwell of University of British Columbia, began work on subjective well-being (SWB) in the 1990s. This work expanded through the World Happiness Report, first published in 2011, and updated annually. In 2025, the World Happiness Report ranked Canada 18th globally (remaining the happiest country in the G7).

Despite this progress, worrisome trends in Canada's life expectancy and SWB have emerged. Canada's level relative to other countries is falling, and the difference between age cohorts is among the highest in the world. This decline is evident in increasing mental health issues in Canada's youth. Between 2011 and 2018, poor perceived mental health among youth aged 12–24 increased from 4.2 to 9.9% [12] (about double for teen girls over boys [13]). Another worrying trend is that Canada experienced a decline in life expectancy in three consecutive years (from 82.3 in 2019 to 81.3 years in 2022 [14]). This decline was largely due to COVID-19, and rebounded slightly in 2023, however increasing levels of addiction, and self-harm continue to contribute to the decline [15]. Opioid-related deaths in Canada average 6000 per year with most deaths involving young males (72%), especially in the cities of Vancouver, Kelowna, Edmonton, Sudbury, Thunder Bay, Hamilton, and Toronto [16].

Canada is one of the few industrialized countries where pedestrian fatalities are increasing, averaging over 300 fatalities annually [17]. In Calgary, for example, pedestrian fatalities increased from 4 in 2023 to 13 in 2024 [18].

For the last 100 years, cities largely focused on improving public health through basic service provisions such as water supply and air quality, reliable electricity, and waste removal. The next 100 years will be far more difficult as these systems need to be expanded and strengthened within a changing climate, but also greater emphasis on more disparate health impacts is needed. These efforts need to be undertaken within changing populations, demographics, and economy. Vision zero (traffic fatalities) with sustained declines, addiction services, community programs for youth, elder support, vector control, and increasing assistance through pandemics and natural disasters; these public health issues are best provided through local governments; however, they are not well-equipped to respond to the growing demands.

A challenge for Canadian cities is that although public health and safety is best strengthened from the ground up, many of the key policy and finance levers are not under local government's remit. For example, local governments may advocate for climate mitigation, yet they have direct control over less than a few percentages of overall GHG emissions. The increase in pedestrian deaths is linked to growing vehicle size in Canada [19]. Municipalities have little say in Canada's shifting vehicle fleet, although the majority of the country's roads are under their jurisdiction.

Local governments need a dynamic, interdisciplinary approach to local public health (and sustainability). Civil engineers, doctors, public health advocates, urban planners, with myriad public policy, financing, and governance practitioners, need to act on behalf of the collective. These individuals tend to be employed across many governments, agencies, and utilities. COVID-19 provided a dress-rehearsal for many of the challenges likely to arise over the next 100 years. Key lessons highlight the value of timely, credible data, provided by, and to, municipalities. Monitoring infection loads through ongoing municipal wastewater testing, providing a powerful example. The benefits of open communication cannot be over-stated.

Civility

Populist politicians like Rob Ford (Toronto councilor and mayor 2000–2016) and Donald Trump, manifest a collective id. Stoking anger and fear, increasingly emboldened through social media, populism and its broader form of nationalism, nurture and respond to an aggrieved segment of society. The ill will of many is not tempered, but rather directed at real and perceived enemies, and those attributed with causing the grievance.

In civilization and its discontent, Sigmund Freud addresses the fundamental paradox of civilization (i.e., cities): civilization emerged to protect us from unhappiness, and yet civilization is our largest source of unhappiness. Civilization at times must compromise individual happiness to fulfill its primary goal of peaceful coexistence.

Canada's cities have largely been understated within the urban dialogue for the last 100 years. Jane Jacobs contributed to the discussion, but only after she and her family moved to Toronto to avoid the Vietnam War. Her urban thoughts grew from New York City. Lewis Mumford's more pessimistic discussion of *The City in History* does not mention any of Canada's larger cities [20]. While Peter Hall's more upbeat *Cities in Civilization* mentions Toronto, but only as contributing through the Reichmann brothers financial support to the Canary Warf development in London [21].

Robert Fishman in reviewing Peter Hall's *Cities in Civilization* highlights the argument that great cities are central to civilization because their very size and complexity make them natural sites of "the innovative milieu". He goes on to suggest "only the greatest cities can bring together the critical mass of creative people to overcome cultural inertia. Within these urban networks of innovators, new paradigms take shape that transform civilization. Here lies the justification and the salvation of the city" [22].

Canada's cities are poised to be among the most influential communities of this century. Like Americans avoiding the Vietnam War, Canada's cities may become home to those eschewing emerging nationalism and curtailment of liberties and possibly growing anger of the USA. They will be seeking a common civility.

Canada's cities will need to build civility between urban and regional voices. The often well-educated and mobile *anywheres* need to live comfortably with the more rooted, traditional *somewheres*. Toronto needs to rejoice in the flourishing of Montreal, Calgary, and Vancouver. And vice versa. And all Canadian cities need to flourish in partnership with their rural hinterlands. Similarly, Toronto needs to grow with Detroit, Chicago, Buffalo, and New York City; Montreal with Plattsburgh, NY, and Burlington, VT; and Vancouver with Seattle and Portland. Geography will continue to bind urban regions across borders; however, civility will ensure a common future.

Pragmatism

Cogitando et Agendo Ducemus, by thinking and doing, we shall lead. Ontario Tech University's motto illustrates that both thought *and* action are necessary to achieve great things. Pragmatism and the strength of cities demonstrate this well. Local government is the most pragmatic level of government. This is well represented in the quip by former New York City Mayor Fiorello LaGuardia that there is no Republican or Democratic way to pick up the garbage. The pragmatism of cities is especially needed as local governments now own and manage more than half the country's infrastructure. Cities will continue to chafe at provincial and federal government initiatives driven more by ideology and piece meal support, e.g., transit development with sporadic provincial and federal funding requirements, removing Toronto's bike lanes (again), federal immigration and foreign student levels and housing initiatives, language laws imposed on Montreal.

Globally, as cities (urban communities) develop a shared voice, a much stronger expression for common sense and pragmatism may emerge. The whims of a national leader that might wipe out wealth overnight and degrade the ability of cities to enhance quality of life, may face stronger, more coordinated opposition. As climate impacts intensify, cities will urge more aggressive mitigation efforts, and likely more pragmatic, and open, burden sharing. Cities are fertile ground for growing discontent with civilization. Correspondingly, although messy and erratic, cities are best positioned to weed out this discontent before it clogs a city's, and a country's, ability to serve the public.

Municipal pooling (Sect. 7.13) is another area where a renewed partnership is needed between residents and businesses and municipalities. Cities need to mediate and consolidate services, monitor data management systems, and serve residents through new public–private partnerships. In 2012, a Toronto taxi medallion cost as much as $360,000. With the arrival of Uber and Lyft, these prices plummeted. Drivers today mostly apply for the $222 vehicle-for-hire driver license. Disruptions like this will continue; however for the benefit of residents and local businesses, municipalities should be more proactive in identifying and encouraging public benefit in these emerging trends.

Municipal pooling also includes shared infrastructure. Canada was knitted together through the Canadian Pacific Railway, and later the TransCanada Highway,

Alouette satellites, and the national air carrier. The next, shared infrastructure, largely to connect Canadian cities, will likely be a national electricity transmission grid, and data management systems and communications platforms, to support local governments and utilities in connecting their communities.

Cities need to think and feel at the same time. Empathy is not a luxury in cities; it is a pragmatic way to encourage civility. People need to believe the streets of the city are welcoming and safe, not only to enhance quality of life and personal well-being, but as a key driver of economic (and social) productivity. Cities, especially in Canada, tend to have higher cultural differences across the community. Cities manifest the concept that 'our diversity is our strength'.

Cities are anchored in the knowledge that they existed before their host country and are likely to exist long after the country is supplanted by the next one.

References

1. https://worldhappiness.report/ (3-29-2025)
2. Black C (2014) Rise to greatness: the history of Canada from the Vikings to the present. McClelland & Stewart
3. See Nate Hagens. https://www.thegreatsimplification.com/
4. Statistics Canada (2024) estimated half of 2019–20 and 2020–21 reference years. 400,058 people born in 2020 expected to eventually immigrate to Canada (author estimate)
5. https://ourworldindata.org/grapher/births-and-deaths-projected-to-2100. Accessed 26 Nov 2024
6. For Canadian babies 64 percent of females, and 54 percent of males, are expected to reach their 90th birthdays (2.3 times and 3.9 times more likely than if she or he, respectively, were born in 1925). Canadian children born in 2019 have an 18% chance of living to 100 years if female, 11% if male
7. Mortality Projections for Social Security Programs in Canada, Actuarial Study No. 22. Office of the Chief Actuary, December 2021
8. Tekaroniake Evans T (2023) How the Iroquois confederacy was formed. https://www.history.com/news/iroquois-confederacy-hiawatha-peacemaker-great-law-of-peace
9. CBC Radio, Ideas; The Shock of the New | The Year 1947: Fractures and Tectonic Shifts (originally aired Nov 2022)
10. Rushdie S (2021) Languages of truth: essays 2003–2020. Random House
11. OurworldinData.com Life Expectancy. Accessed 30 Mar 2025
12. Wiens K, Bhattarai A, Pedram P, Dores A, Williams J, Bulloch A, Patten S (2020) A growing need for youth mental health services in Canada: examining trends in youth mental health from 2011 to 2018. Epidemiol Psychiatr Sci 29:e115
13. Statistics Canada (2024-09-10), 2023 Canadian Health Survey on Children and Youth
14. Statistics Canada (2023-11-27), Deaths, 2022
15. Statistics Canada (2024-12-04), Deaths, 2023
16. Government of Canada (2025-03-07) Opioid- and Stimulant-related harms in Canada
17. Statistics Canada (2023-10-30), Circumstances surrounding pedestrian fatalities, 2018 to 2020.
18. Calgary Herald, April 1, 2025. Steven Wilhelm and Noah Brennan
19. Canadian Association of Road Safety Professionals (2024) Bigger, heavier, higher…large SUVs and pickups pose serious safety threats to vulnerable road users
20. Mumford L (1961) The city in history: its origins, its transformations, and its prospects. Houghton Mifflin Harcourt
21. Hall PG (1998) Cities in civilization. Pantheon Books, New York

22. Fishman R (1999) Cities in civilization, review, Fall ed. Harvard Design Magazine

Open Access This chapter is licensed under the terms of the Creative Commons Attribution-NonCommercial-NoDerivatives 4.0 International License (http://creativecommons.org/licenses/by-nc-nd/4.0/), which permits any noncommercial use, sharing, distribution and reproduction in any medium or format, as long as you give appropriate credit to the original author(s) and the source, provide a link to the Creative Commons license and indicate if you modified the licensed material. You do not have permission under this license to share adapted material derived from this chapter or parts of it.

The images or other third party material in this chapter are included in the chapter's Creative Commons license, unless indicated otherwise in a credit line to the material. If material is not included in the chapter's Creative Commons license and your intended use is not permitted by statutory regulation or exceeds the permitted use, you will need to obtain permission directly from the copyright holder.

MIX
Papier aus verantwortungsvollen Quellen
Paper from responsible sources
FSC® C105338

If you have any concerns about our products,
you can contact us on
ProductSafety@springernature.com

In case Publisher is established outside the EU,
the EU authorized representative is:
**Springer Nature Customer Service Center GmbH
Europaplatz 3, 69115 Heidelberg, Germany**

Printed by Libri Plureos GmbH
in Hamburg, Germany